Dimensions Mc
Textbook 3B

MW01492531

Authors and Reviewers

Bill Jackson

Jenny Kempe

Cassandra Turner

Allison Coates

Tricia Salerno

Consultant

Dr. Richard Askey

Singapore Math Inc.

Published by Singapore Math Inc.

19535 SW 129th Avenue

Tualatin, OR 97062

www.singaporemath.com

Dimensions Math® Textbook 3B

ISBN 978-1-947226-09-8

First published 2018

Reprinted 2019, 2020 (twice), 2021

Printed in China

Acknowledgments

Editing by the Singapore Math Inc. team.

Design and illustration by Cameron Wray with Carli Bartlett.

Preface

The Dimensions Math® Pre-Kindergarten to Grade 5 series is based on the pedagogy and methodology of math education in Singapore. The curriculum develops concepts in increasing levels of abstraction, emphasizing the three pedagogical stages: Concrete, Pictorial, and Abstract. Each topic is introduced, then thoughtfully developed through the use of problem solving, student discourse, and opportunities for mastery of skills.

Features and Lesson Components

Students work through the lessons with the help of five friends: Emma, Alex, Sofia, Dion, and Mei. The characters appear throughout the series and help students develop metacognitive reasoning through questions, hints, and ideas.

The colored boxes ▉ and blank lines in the textbook lessons are used to facilitate student discussion. Rather than writing in the textbooks, students can use whiteboards or notebooks to record their ideas, methods, and solutions.

Chapter Opener

Each chapter begins with an engaging scenario that stimulates student curiosity in new concepts. This scenario also provides teachers an opportunity to review skills.

Think

Students, with guidance from teachers, solve a problem using a variety of methods.

Learn

One or more solutions to the problem in **Think** are presented, along with definitions and other information to consolidate the concepts introduced in **Think**.

Do

A variety of practice problems allow teachers to lead discussion or encourage independent mastery. These activities solidify and deepen student understanding of the concepts.

Exercise

A pencil icon ✏️ at the end of the lesson links to additional practice problems in the workbook.

Practice

Periodic practice provides teachers with opportunities for consolidation, remediation, and assessment.

Review

Cumulative reviews provide ongoing practice of concepts and skills.

Emma   Alex   Sofia   Dion   Mei

Contents

Chapter		Lesson	Page

Chapter		Lesson	Page

Chapter	Lesson	Page

Chapter 8

Multiplying and Dividing with 6, 7, 8, and 9

1 × 6	1 × 7	1 × 8	1 × 9
2 × 6	2 × 7	2 × 8	2 × 9
3 × 6	3 × 7	3 × 8	3 × 9
4 × 6	4 × 7	4 × 8	4 × 9
5 × 6	5 × 7	5 × 8	5 × 9
6 × 6	6 × 7	6 × 8	6 × 9
7 × 6	7 × 7	7 × 8	7 × 9
8 × 6	8 × 7	8 × 8	8 × 9
9 × 6	9 × 7	9 × 8	9 × 9
10 × 6	10 × 7	10 × 8	10 × 9

Which facts do we know already?
Which ones do we still need to learn?

How can we use the facts we know to find the facts we still need to learn?

Think

There are 6 paper towel rolls in each pack.

$1 \times 6 = 6$

How many rolls are there in ⬚2 packs?

⬚2 × 6 = ?

Find the number of rolls if there are...

⬚3 , ⬚4 , ⬚5 , ⬚6 , ⬚7 , ⬚8 , ⬚9 , and ⬚10 packs.

How can we use the facts for 2, 3, 4, 5, and 10 to find any new facts for 6?

Do you notice any patterns?

Learn

$1 \times 6 = 6$

$+1$ $+6$ $+6$

$2 \times 6 = 12$

$3 \times 6 = 18$

$4 \times 6 = 24$

$5 \times 6 = 30$

$6 \times 6 = 36$ 6×6 is [   ] more than 5×6.

$7 \times 6 = 42$

$8 \times 6 = 48$

$9 \times 6 = 54$ $9 \times 6 = $ [   ] $\times 6 - 6$

$10 \times 6 = 60$

I noticed a pattern in the sum of the digits in the products.
12: 1 + 2 = 3
18: 1 + 8 = 9
24: 2 + 4 = 6
30: 3 + 0 = 3
...

I noticed that the products are all even numbers.

Do

1

$5 \times 6 = 30$

$1 \times 6 =$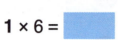

$6 \times 6 =$

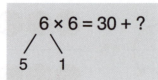

$6 \times 6 = 30 + ?$

$5 \quad 1$

2

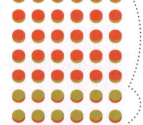

$5 \times 6 = 30$

$2 \times 6 =$

$7 \times 6 =$

$7 \times 6 = 30 + ?$

$5 \quad 2$

3

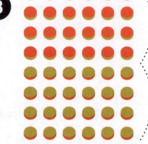

$4 \times 6 = 24$

$4 \times 6 = 24$

$8 \times 6 =$

$8 \times 6 = 2 \times 24$

4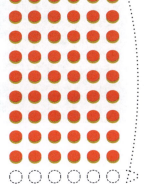

$10 \times 6 = 60$

$9 \times 6 = 60 - ?$

$1 \times 6 =$

$9 \times 6 =$

5

$8 \times 3 =$

$8 \times 3 =$

$8 \times 6 =$

The multiplication facts for 6 are all double the facts for 3.

---

6 (a) $6 \times 6 =$

(b) $7 \times 6 =$ | $6 \times 7 =$

(c) $8 \times 6 =$ | $6 \times 8 =$

(d) $9 \times 6 =$ | $6 \times 9 =$

7 5 × 6 = [ ] | [ ] ÷ 5 = 6

6 × 5 = [ ] | [ ] ÷ 6 = 5

8 (a) [ ] × 6 = 42 | 42 ÷ 6 = [ ]

(b) [ ] × 6 = 54 | 54 ÷ 6 = [ ]

(c) [ ] × 6 = 48 | 48 ÷ 6 = [ ]

(d) [ ] × 6 = 36 | 36 ÷ 6 = [ ]

9 Find the quotient and remainder.

(a) 24 ÷ 6 | 28 ÷ 6

(b) 36 ÷ 6 | 41 ÷ 6

(c) 54 ÷ 6 | 57 ÷ 6

What is the greatest remainder
we can have when dividing by 6?

Think

These painting canvases come in packs of 7.

$1 \times 7 = 7$

How many canvases are there in ⬚2 packs?

⬚2 $\times 7 = ?$

Find the number of canvases if there are...

⬚3 , ⬚4 , ⬚5 , ⬚6 , ⬚7 , ⬚8 , ⬚9 , and ⬚10 packs.

How can we use the facts we know to find any facts for 7 we have not learned yet?

Learn

$1 \times 7 = 7$

+1

$2 \times 7 = 14$ **+7**

$3 \times 7 = 21$

$4 \times 7 = 28$

$5 \times 7 = 35$

$6 \times 7 = 42$

$7 \times 7 = 49$ $7 \times 7 = 6 \times 7 +$

$8 \times 7 = 56$ $8 \times 7 = 7 \times 7 +$

$9 \times 7 = 63$ $9 \times 7 = 10 \times 7 -$

$10 \times 7 = 70$

+7

There are only 3 new facts.

Do

1

$3 \times 7 = $

$3 \times 7 = $

$6 \times 7 = $

I can double the answer to 3 × 7 to find 6 × 7.

2

$5 \times 7 = $

$2 \times 7 = $

$7 \times 7 = $

$7 \times 7 = 35 + ?$

5   2

3

$8 \times 5 = 40$

$8 \times 2 = $

$8 \times 7 = $

$8 \times 7 = 40 + ?$

5   2

6 tens 4 ones
74

74 − 10
64

4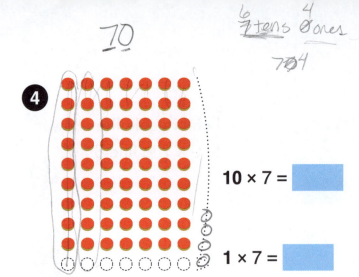

10 × 7 =

1 × 7 =

9 × 7 is 7 less than 10 × 7.

9 × 7 =

---

5 (a) 6 × 7 =   7 × 6 =

(b) 7 × 7 =

(c) 8 × 7 =   7 × 8 =

(d) 9 × 7 =   7 × 9 =

6 6 × 7 =   ÷ 7 = 6

7 × 6 =   ÷ 6 = 7

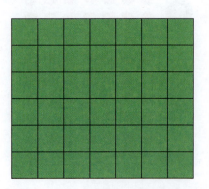

7 (a) [ ] × 7 = 63　│　63 ÷ 7 = [ ]

(b) [ ] × 7 = 49　│　49 ÷ 7 = [ ]

(c) [ ] × 7 = 56　│　56 ÷ 7 = [ ]

8 Paul paid $28 for 7 identical tote bags. How much did each tote bag cost?

9 Elena's cat weighs 7 pounds. Her dog is 9 times as heavy as her cat. How much does her dog weigh?

10 Find the quotient and remainder.

(a) 21 ÷ 7　│　24 ÷ 7

(b) 49 ÷ 7　│　50 ÷ 7

(c) 63 ÷ 7　│　69 ÷ 7

What is the greatest remainder we can have when dividing by 7?

Exercise 2 • page 4

Think

There are 275 bandages in one pack.
How many bandages are in 6 packs?

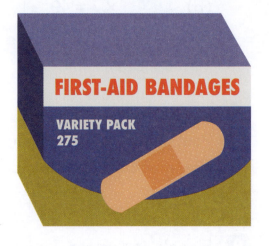

FIRST-AID BANDAGES

VARIETY PACK
275

Learn

275 × 6

```
      3
    2 7 5
×       6
_____
        0
```
Multiply 5 ones by 6.
5 ones × 6 = **3** tens **0** ones

↓

```
    4 3
    2 7 5
×       6
_____
      5 0
```
Multiply 7 tens by 6 and add any regrouped tens.
(7 tens × 6) + 3 tens = **4** hundreds **5** tens

↓

```
    4 3
    2 7 5
×       6
_____
  1,6 5 0
```
Multiply 2 hundreds by 6 and add any regrouped hundreds.
(2 hundreds × 6) + 4 hundreds
= **1** thousand **6** hundreds

There are ▢ bandages in 6 packs.

Do

1 Find the product of 69 and 6.

$$
\begin{array}{r}
6\,9 \\
\times 6 \\
\hline
\end{array}
$$

2 Multiply 867 by 7.

$$
\begin{array}{r}
8\,6\,7 \\
\times 7 \\
\hline
\end{array}
$$

3 Find the product.

(a) 81 × 6   (b) 16 × 7   (c) 7 × 68

(d) 208 × 7   (e) 7 × 670   (f) 468 × 6

4 An all-day ticket to the amusement park costs $126.
How much will 6 tickets cost?

5 A sports store sold 258 baseball caps for $7 each.
How much money did it receive from the sale of the caps?

Exercise 3 • page 7

Think

956 eggs were put into cartons that hold 6 eggs each.
How many cartons were used?
Are there any eggs left over?

Learn

```
        1
   6 ) 9  5  6
       6
       3
```

Divide 9 hundreds by 6.
9 hundreds ÷ 6 is **1** hundred with
3 hundreds left over.

```
        1  5
   6 ) 9  5  6
       6
       3  5
       3  0
          5
```

Regroup the remaining hundreds into tens.
Divide the 35 tens by 6.
35 tens ÷ 6 is **5** tens with
5 tens left over.

```
        1  5  9
   6 ) 9  5  6
       6
       3  5
       3  0
          5  6
          5  4
             2
```

Regroup the remaining tens into ones.
Divide 56 ones by 6.
56 ones ÷ 6 is **9** ones with
2 left over.

 cartons were used and [   ] eggs were left over.

Do

1 (a) Divide 98 by 6.

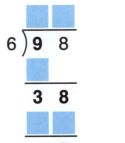

6)9 8   9 tens ÷ 6

3 8   38 ones ÷ 6

2

The remainder should be less than...

(b) Divide 896 by 7.

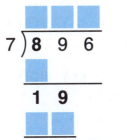

7)8 9 6   8 hundreds ÷ 7

1 9   19 tens ÷ 7

5 6   56 ones ÷ 7

0

(c) Divide 732 by 7.

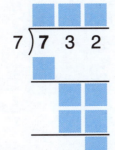

7)7 3 2   7 hundreds ÷ 7

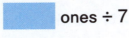 ones ÷ 7

Remember to check your answers.
104 × 7 + 4 = 732

(d) Divide 474 by 6.

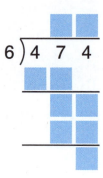

$$6\overline{)4\ 7\ 4}$$

2 Find the quotient and remainder.

(a) $84 \div 6$ (b) $95 \div 7$ (c) $709 \div 6$

(d) $916 \div 7$ (e) $500 \div 6$ (f) $800 \div 7$

3

Mr. Kalani drove 750 miles in 6 days.
He drove the same distance each day.
How far did he drive each day?

4 A gardener spent $644 on pots.
Each pot cost $7.
How many pots did she buy?

Exercise 4 • page 10

1 Find the value.

(a) 6 × 7

(b) 8 × 7

(c) 6 × 8

(d) 63 ÷ 7

(e) 49 ÷ 7

(f) 54 ÷ 6

2 Solve.

(a) 64 × 7

(b) 563 × 6

(c) 705 × 6

(d) 92 ÷ 6

(e) 920 ÷ 7

(f) 302 ÷ 7

3 Rodrigo bought 8 succulents for $7 each.
He gave $60 to the cashier.
How much change did he get back?

4 At soccer practice, June ran the length of a 120-m soccer field 7 times.
Then she ran the length of the field another 6 times.
How far did she run in all?

5 259 days of the year have passed.
There are 52 weeks in the year.
How many weeks are left?

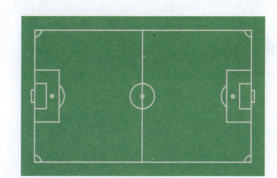

Exercise 5 • page 13

Think

These bottles of juice come in packs of 8.

$1 \times 8 = 8$

How many bottles of juice are there in ⟦2⟧ packs?

⟦2⟧ $\times 8 = ?$

Find the number of bottles of juice if there are...

⟦3⟧ , ⟦4⟧ , ⟦5⟧ , ⟦6⟧ , ⟦7⟧ , ⟦8⟧ , ⟦9⟧ , and ⟦10⟧ packs.

How can we use the facts we know to find any facts for 8 we have not learned yet?

Learn

$1 \times 8 = 8$

+1 +8 +8

$2 \times 8 = 16$

$3 \times 8 = 24$

$4 \times 8 = 32$

$5 \times 8 = 40$

$6 \times 8 = 48$

$7 \times 8 = 56$　　　　$7 \times 8 = 6 \times 8 +$ ▮

$8 \times 8 = 64$　　　　$8 \times 8 = 7 \times 8 +$ ▮

$9 \times 8 = 72$　　　　$9 \times 8 = 10 \times 8 -$ ▮

$10 \times 8 = 80$

There are only 2 new facts.

I noticed that the ones digit has a pattern of 8, 6, 4, 2, 0. They are all even numbers.

Do

1

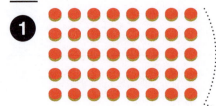

$5 \times 8 = 40$

$1 \times 8 =$

$6 \times 8 =$

2

$5 \times 8 = 40$

$2 \times 8 =$

$7 \times 8 =$

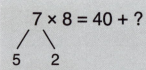

$7 \times 8 = 40 + ?$

5   2

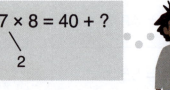

3

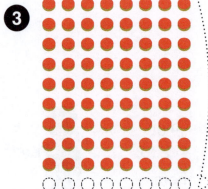

$10 \times 8 =$

$1 \times 8 =$

$9 \times 8 =$

×2 **×2**

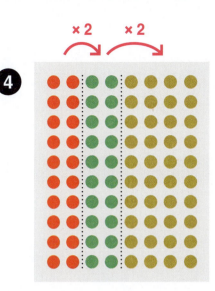

4

×2　　　　　　　　　　　×2

2 × 2 = 4	4 × 2 = 8	8 × 2 = 16
2 × 3 = 6	4 × 3 = 12	8 × 3 = 24
2 × 4 = 8	4 × 4 = 16	8 × 4 = 32
2 × 5 = 10	4 × 5 = ▢	8 × 5 = ▢
2 × 6 = 12	4 × 6 = ▢	8 × 6 = ▢
2 × 7 = 14	4 × 7 = ▢	8 × 7 = ▢
2 × 8 = 16	4 × 8 = ▢	8 × 8 = ▢
2 × 9 = 18	4 × 9 = ▢	8 × 9 = ▢
2 × 10 = ▢	4 × 10 = ▢	8 × 10 = ▢

5 $7 \times 8 =$ ▢ | ▢ $\div 7 = 8$

$8 \times 7 =$ ▢ | ▢ $\div 8 = 7$

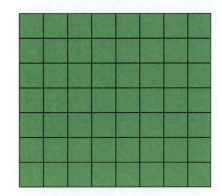

6 (a) ▢ $\times 8 = 64$ | $64 \div 8 =$ ▢

(b) ▢ $\times 8 = 72$ | $72 \div 8 =$ ▢

7 Find the quotient and remainder.

(a) $48 \div 8$ | $51 \div 8$

(b) $32 \div 8$ | $36 \div 8$

(c) $64 \div 8$ | $70 \div 8$

(d) $56 \div 8$ | $63 \div 8$

8 (a) ▢ $\times 8 + 2 = 74$

(b) ▢ $\times 8 + 6 = 54$

Exercise 6 • page 17

8-6 The Multiplication Table of 8

Think

These light bulbs come in packs of 9.

$1 \times 9 = 9$

How many light bulbs are there in ☐2☐ packs?

☐2☐ $\times 9 = ?$

Find the number of light bulbs if there are...

☐3☐, ☐4☐, ☐5☐, ☐6☐, ☐7☐, ☐8☐, ☐9☐, and ☐10☐ packs.

Do you notice any patterns?

There is only 1 new fact!

Learn

+1 $1 \times 9 = 9$ +9

$2 \times 9 = 18$ +9

$3 \times 9 = 27$

$4 \times 9 = 36$

$5 \times 9 = 45$

$6 \times 9 = 54$

$7 \times 9 = 63$ $7 \times 9 = 6 \times 9 +$

$8 \times 9 = 72$

$9 \times 9 = 81$ $9 \times 9 = 10 \times 9 -$

$10 \times 9 = 90$

The tens digit is 1 less...

$6 \times 9 = 54$
$7 \times 9 = 63$
$8 \times 9 = 72$
...

$5 + 4 = 9$
$6 + 3 = 9$
$7 + 2 = 9$
If I add the digits in the product, the sum is 9.

Do

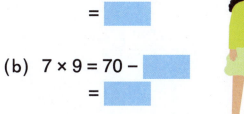
I can use the multiplication facts for 10 to find the facts for 9.

(a) $6 \times 9 = 60 - 6$

         =

(b) $7 \times 9 = 70 - $

         =

(c) $8 \times 9 = 80 - $

         =

(d) $9 \times 9 = 90 - $

         =

2 (a) $5 \times 9 = 45$

         $6 \times 9 = 45 + 10 - 1$

                =

(b) $8 \times 9 = $

(c) $9 \times 9 = $

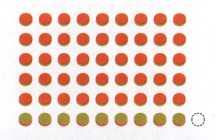

$6 \times 9 = 5\ 4$

$+\ 1\ ten \quad -\ 1$

$7 \times 9 = 6\ 3$

3 (a) ☐ × 9 = 81 | 81 ÷ 9 = ☐

(b) ☐ × 9 = 63 | 63 ÷ 9 = ☐

4 Find the quotient and remainder.

(a) 36 ÷ 9 | 44 ÷ 9

(b) 63 ÷ 9 | 70 ÷ 9

(c) 45 ÷ 9 | 48 ÷ 9

5 A 10-person van can hold 1 driver and 9 passengers.
What is the least number of drivers needed for a group of 60 passengers?

6 For a game, 9 children are trying to share 85 tokens equally.
Each child will get as many tokens as possible.
How many tokens will be shared?
How many tokens will be left over?

Exercise 7 • page 20

8-7 The Multiplication Table of 9

Think

This tablet costs $976.

How much will it cost to buy 8 of these tablets?

Learn

976 × 8

```
      4
  9 7 6
×     8
──────
      8
```

Multiply 6 ones by 8.

6 ones × 8 = **4** tens **8** ones

↓

```
  6 4
  9 7 6
×     8
──────
    0 8
```

Multiply 7 tens by 8 and add any regrouped tens.

(7 tens × 8) + 4 tens = **6** hundreds **0** tens

↓

```
  6 4
  9 7 6
×     8
──────
7,8 0 8
```

Multiply 9 hundreds by 8 and add any regrouped hundreds.

(9 hundreds × 8) + 6 hundreds

= **7** thousands **8** hundreds

It will cost $ _____ .

Do

1 Multiply 76 by 9.

```
     76
  ×   9
  ____
```

2 Find the product of 8 and 498.

```
    498
  ×   8
  ____
```

3 Find the product.

(a) 71 × 9   (b) 17 × 8   (c) 9 × 85

(d) 309 × 8   (e) 9 × 680   (f) 742 × 9

4 There are 150 yo-yos in a box.
How many yo-yos are in 9 such boxes?

5 A table costs 9 times as much as a chair.
The chair costs $99.
How much more does the table cost than the chair?

Exercise 8 • page 23

Think

954 donuts were put into cartons.
Each carton holds 9 donuts.
How many cartons were used?
Were there any donuts left over?

Learn

```
     1
  9)9 5 4       Divide 9 hundreds by 9.
    9           9 hundreds ÷ 9 = 1 hundred
  ─────
    0
```

```
     1 0
  9)9 5 4
    9
  ─────
      5         There are not enough tens to divide tens by 9.
```

```
     1 0 6
  9)9 5 4
    9
  ─────
      5 4       Regroup the tens into ones.
      5 4       Divide the 54 ones by 9.
  ─────────
        0       54 ones ÷ 9 = 6 ones
```

 cartons were used. | donuts were left over.

Do

1 (a) Divide 97 by 8.

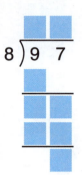

$$8\overline{)9\ 7}$$

(b) Divide 996 by 8.

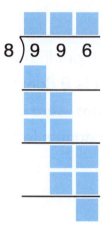

$$8\overline{)9\ 9\ 6}$$

(c) Divide 388 by 9.

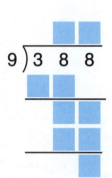

$$9\overline{)3\ 8\ 8}$$

2 Find the quotient and remainder.

(a) $99 \div 9$

(b) $979 \div 8$

(c) $809 \div 8$

(d) $753 \div 9$

(e) $500 \div 8$

(f) $800 \div 9$

3 244 children are making teams of 8 children for a tug-of-war competition.
They want as many teams as possible.
The children not on a team will be the referees.
How many referees will there be?

Exercise 9 • page 26

8-9 Dividing by 8 and 9

1 (a) $9 \times \boxed{} = 81$

(b) $\boxed{} \times 9 = 72$

(c) $64 \div 8 = \boxed{}$

(d) $\boxed{} \div 8 = 7$

(e) $\boxed{} = 7 \times 9 + 4$

(f) $53 = \boxed{} \times 8 + 5$

2 Find the product.

(a) 700×8

(b) 9×60

(c) 66×8

(d) 9×83

(e) 9×697

(f) 745×8

3 Find the quotient and remainder.

(a) $62 \div 8$

(b) $30 \div 9$

(c) $100 \div 8$

(d) $924 \div 8$

(e) $379 \div 9$

(f) $500 \div 9$

4 Diego saved $18.

He saved twice as much money as Taylor.

Yara saved 9 times as much money as Taylor.

How much less money did Taylor save than Yara?

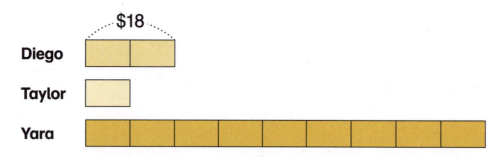

5 A bicycle costs $342.

A motorcycle costs 8 times as much as a bicycle.

How much do the motorcycle and bicycle cost altogether?

6 A store employee put 448 apples into bags of 8 apples each. The store sold all the bags of apples for $9 per bag. How much money did it receive from the sale of the apples?

7 There are 396 children in an elementary school.

There are 9 times as many children in the elementary school as there are in the preschool.

How many more children are in the elementary school than in the preschool?

Exercise 10 • page 29

8-10 Practice B

Chapter 9

Fractions — Part 1

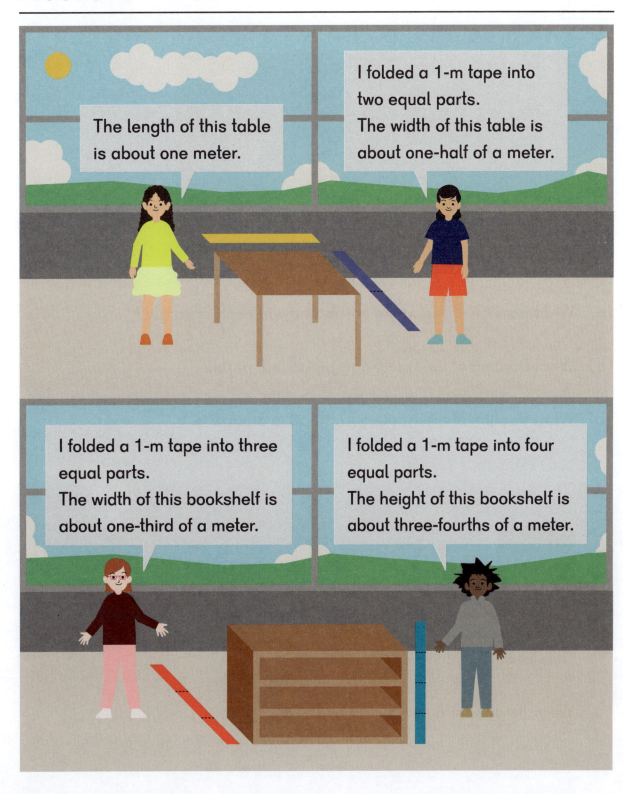

Think

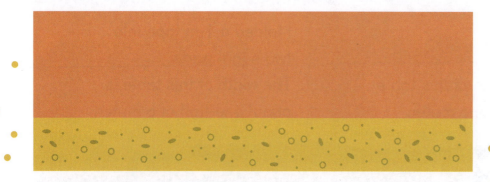

3 children are sharing this cornbread equally.

(a) What fraction of the whole cornbread will each child get?

(b) What fraction of the whole cornbread is 2 parts?

(c) What fraction of the whole cornbread is 3 parts?

Learn

(a)

When we divide 1 whole into 3 equal parts, each part is one-third.

We write one-third as $\frac{1}{3}$.

(b)

When we divide 1 whole into 3 equal parts, 2 parts is two-thirds.

We write two-thirds as $\frac{2}{3}$.

There are ▢ one-thirds in $\frac{2}{3}$.

(c)

When we divide 1 whole into 3 equal parts, 3 parts is three-thirds.

We write three-thirds as $\frac{3}{3}$.

There are ▢ one-thirds in $\frac{3}{3}$.

$\frac{3}{3}$ is equal to 1.

Fractions are numbers that count <u>equally</u> divided parts of the whole.

$\dfrac{2}{3}$ ← numerator
← denominator

The **denominator** tells us how many equal parts the whole is divided into.
The **numerator** counts the number of parts.

Do

1 Identify the numerator and the denominator in each fraction.

(a) $\frac{1}{4}$

(b) $\frac{7}{12}$

(c) $\frac{5}{6}$

(d) $\frac{5}{9}$

2 (a) $\frac{\square}{4}$ of the circle is colored.

(b) $\frac{\square}{4}$ of the circle is not colored.

(c) $1 = \frac{\square}{4}$ 1 whole = 4 fourths

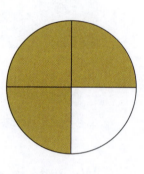

(d) $\frac{\square}{4}$ and $\frac{1}{4}$ make 1.

3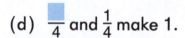

(a) $\frac{\square}{\square}$ of the bar is colored.

(b) $\frac{\square}{\square}$ of the bar is not colored.

(c) 1 whole = $\square$ sevenths

(d) $1 = \frac{\square}{7}$

(e) $\frac{5}{7}$ and $\frac{\square}{\square}$ make 1.

4 What fraction of each shape is colored?
What fraction is not colored?

(a)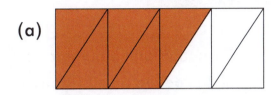

(b)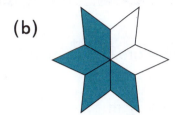

(c)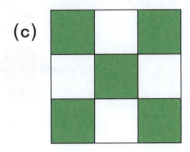

(d)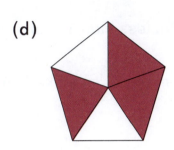

5 (a) $\frac{7}{8}$ and $\frac{}{}$ make 1.

(b) $\frac{5}{12}$ and $\frac{}{}$ make 1.

6 (a) What fraction of a liter of water is in this beaker?

(b) How many $\frac{1}{5}$ L are in $\frac{3}{5}$ L?

(c) How much more water should be added to this beaker to make 1 L?

(d) $\frac{3}{5}$ L and $\frac{}{}$ L make 1 L.

7

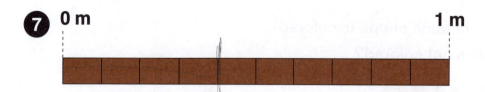

0 m   1 m

(a) There are ☐ $\frac{1}{10}$ m in 1 m.

(b) 1 m = $\frac{\square}{10}$ m

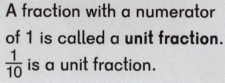

A fraction with a numerator of 1 is called a **unit fraction**. $\frac{1}{10}$ is a unit fraction.

(c) $\frac{1}{10}$ m less than 1 m is $\frac{\square}{\square}$ m.

(d) $\frac{1}{10}$ m and $\frac{\square}{\square}$ m make 1 m.

8 A yellow ribbon is $\frac{2}{3}$ m long.

A green ribbon is $\frac{2}{3}$ ft long.

0 m   1 m

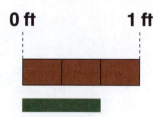

0 ft   1 ft

Is $\frac{2}{3}$ m the same as $\frac{2}{3}$ ft?

Explain why or why not.

Exercise 1 • page 33

Lesson 2
Fractions on a Number Line

Think

A paper tape is marked so that there are 5 equal parts between each meter. The tape is used to measure the length of two ribbons.

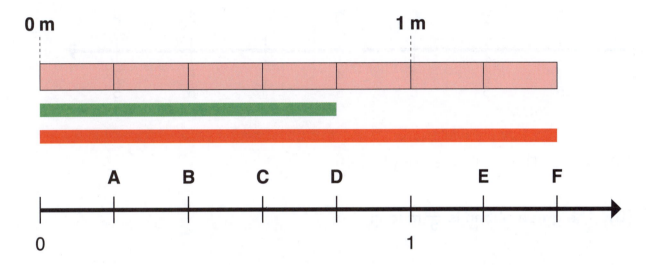

(a) What numbers are indicated by each letter, A, B, C, and D, on the number line?

(b) How long is the green ribbon?

(c) How many $\frac{1}{5}$ m are in 1 m?

(d) How many fifths make 1?

(e) How many fifths make the numbers indicated by E and F?

(f) What numbers are indicated by the letters E and F?

(g) How long is the red ribbon?

Learn

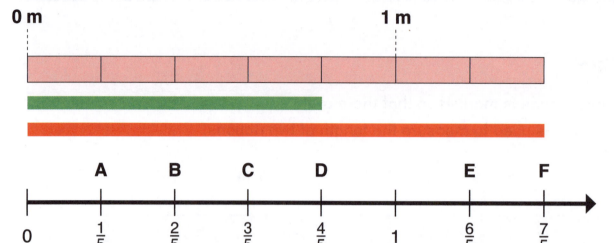

(a) A = $\frac{\blacksquare}{5}$, B = $\frac{\blacksquare}{5}$, C = $\frac{\blacksquare}{5}$, and D = $\frac{\blacksquare}{5}$.

(b) The green ribbon is $\frac{\blacksquare}{5}$ m long.

(c) There are five $\frac{1}{5}$ m in 1 m.

> We can think of a number line as if it were a ruler and divide the distance between 0 and 1 into equal-length parts.

(d) $\blacksquare$ fifths make 1.

(e) E is $\blacksquare$ fifths and F is $\blacksquare$ fifths.

(f) E is $\frac{\blacksquare}{5}$ and F is $\frac{\blacksquare}{5}$.

> Fractions are numbers. They can be represented on a number line.

(g) The red ribbon is $\frac{\blacksquare}{5}$ m long.

Do

1

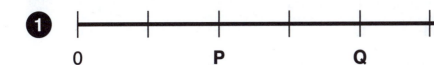

0   P   Q   R   1

(a) What numbers are indicated by each letter, P, Q, and R, on this number line?

(b) Write 1 using sevenths.

(c) R is greater than Q by how much?

(d) P is less than 1 by how much?

(e) What fraction do we add to $\frac{3}{7}$ to get 1?

2

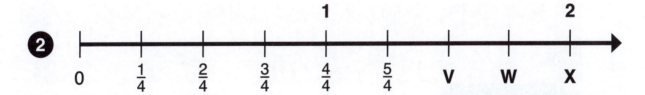

1   2

0   $\frac{1}{4}$ $\frac{2}{4}$ $\frac{3}{4}$ $\frac{4}{4}$ $\frac{5}{4}$ V   W   X

(a) How many fourths make $\frac{5}{4}$?

$\frac{4}{4} = 1$
4 one-fourths make 1.

(b) Write fractions in fourths for V, W, and X.

(c) Which of the fractions marked on the number line are greater than 1 but less than 2?

(d) How many fourths make 2?

3

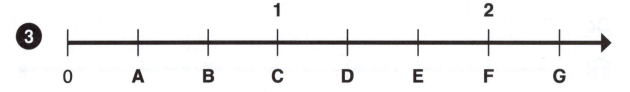

(a) The increment between each tick mark on this number line is .

(b) Write the numbers indicated by each letter.

How do you know if a fraction is greater than 1?

4

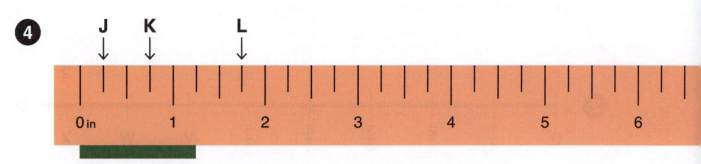

(a) What fractions are indicated by the letters?

(b) Write the length of the green line as a fraction.

5

(a) Which tick marks indicate halves?

(b) What fractions do the other tick marks indicate?

Exercise 2 • page 37

9-2 Fractions on a Number Line

Think

The friends are painting benches.

Mei has $\frac{3}{5}$ gallons of paint left.

Sofia has $\frac{1}{5}$ gallon of paint left.

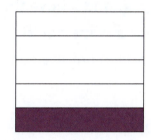

Dion has $\frac{4}{5}$ gallons of paint left.

Alex has $\frac{2}{5}$ gallons of paint left.

How can we put these fractions in order from least to greatest?

Who has the most paint left?

Who has the least paint left?

Can we compare the fractions by comparing their numerators and denominators?

Learn

Sofia	Alex	Mei	Dion

$$\frac{1}{5} \qquad \frac{2}{5} \qquad \frac{3}{5} \qquad \frac{4}{5}$$

The fractions $\frac{1}{5}$, $\frac{2}{5}$, $\frac{3}{5}$, and $\frac{4}{5}$ have the same denominators.
The size of the parts is the same.
The number of parts is not the same.

_____ has the most paint left.

_____ has the least paint left.

If the denominators are the same, we can compare the numerators.

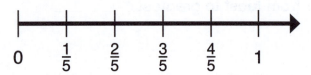

$$0 \qquad \frac{1}{5} \qquad \frac{2}{5} \qquad \frac{3}{5} \qquad \frac{4}{5} \qquad 1$$

On this number line, it is easy to compare the fractions for fifths.

Do

1 Which is greater, $\frac{3}{8}$ or $\frac{5}{8}$?

2 Which is less, $\frac{3}{5}$ or $\frac{2}{5}$?

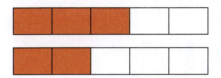

3 Put the fractions in order from least to greatest.

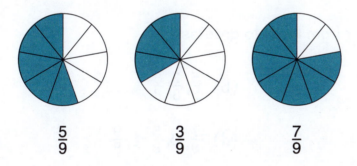

$\frac{5}{9}$ $\qquad$ $\frac{3}{9}$ $\qquad$ $\frac{7}{9}$

4 (a) Locate $\frac{4}{5}$ and $\frac{7}{5}$ on the number line.

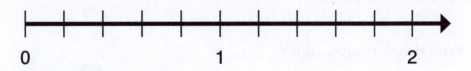

0 $\qquad$ 1 $\qquad$ 2

(b) Which is greater, $\frac{4}{5}$ or $\frac{7}{5}$?

(c) Which is less, 1 or $\frac{7}{5}$?

$1 = \frac{5}{5}$

5 Which is greater, $\frac{3}{7}$ or $\frac{5}{7}$?

6 Which is less, $\frac{17}{12}$ or $\frac{11}{12}$?

7 What sign, >, <, or =, goes in the ◯?

(a) $\frac{1}{4}$ ◯ $\frac{3}{4}$ (b) $\frac{8}{8}$ ◯ 1

(c) $\frac{3}{4}$ ◯ $\frac{6}{4}$ (d) $\frac{6}{9}$ ◯ $\frac{5}{9}$

(e) $\frac{10}{8}$ ◯ $\frac{3}{8}$ (f) $\frac{7}{16}$ ◯ $\frac{5}{16}$

8 Put the numbers in order from least to greatest.

(a) 1, 0, $\frac{1}{2}$ (b) $\frac{5}{8}$, $\frac{3}{8}$, $\frac{6}{8}$

(c) $\frac{2}{6}$, $\frac{5}{6}$, $\frac{8}{6}$, $\frac{3}{6}$ (d) $\frac{7}{8}$, 2, $\frac{5}{8}$, 1, $\frac{9}{8}$

9 Wainani drank $\frac{7}{10}$ L of water at school today.
Olivia drank 1 L of water.

(a) Who drank more water?

(b) How much more water?

Think

The friends are pouring water into containers.

Emma has $\frac{3}{5}$ L of water.

Mei has $\frac{3}{10}$ L of water.

Dion has $\frac{3}{8}$ L of water.

 Can we compare fractions if the denominators are different?

How can we put these fractions in order from least to greatest?

Who has the most water?

Who has the least water?

Learn

Mei	Dion	Emma
$\frac{3}{10}$	$\frac{3}{8}$	$\frac{3}{5}$

The fractions $\frac{3}{10}$, $\frac{3}{8}$, and $\frac{3}{5}$ have the same numerators.
The number of the parts is the same.
The size of each part is not the same.

_____ has the most water.

_____ has the least water.

Compare the denominators in the different fractions.
What do you notice?

Do

1 Which is greater, $\frac{2}{3}$ or $\frac{2}{5}$?

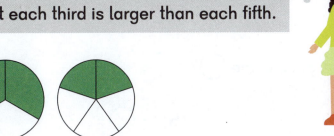

There are the same number of parts, but each third is larger than each fifth.

2 Which is less, $\frac{5}{7}$ or $\frac{5}{9}$?

Which unit fraction is less, $\frac{1}{7}$ or $\frac{1}{9}$?

3 Locate each fraction on the number line next to it.
Then put the fractions in order from least to greatest.

$\frac{4}{8}$

$\frac{4}{10}$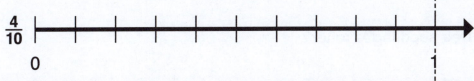

$\frac{4}{6}$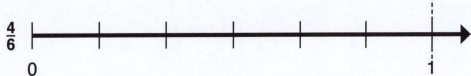

4 Which is greater, $\frac{3}{4}$ or $\frac{3}{5}$?

5 Which is less, $\frac{11}{12}$ or $\frac{11}{16}$?

6 What sign, >, <, or =, goes in the ◯?

(a) $\frac{1}{8}$ ◯ $\frac{1}{3}$

(b) $\frac{2}{3}$ ◯ $\frac{2}{5}$

(c) $\frac{6}{9}$ ◯ $\frac{6}{8}$

(d) $\frac{5}{7}$ ◯ $\frac{5}{8}$

(e) $\frac{10}{8}$ ◯ $\frac{10}{7}$

(f) $\frac{9}{12}$ ◯ $\frac{9}{10}$

7 Put the numbers in order from least to greatest.

(a) $\frac{1}{3}, \frac{1}{8}, \frac{1}{6}$

(b) $\frac{7}{5}, \frac{7}{8}, \frac{7}{6}$

(c) $\frac{3}{4}, \frac{3}{7}, \frac{3}{5}, \frac{3}{8}$

(d) $\frac{2}{5}, 0, \frac{2}{8}, 1$

8 A pine seedling is $\frac{7}{10}$ m tall.

A birch seedling is $\frac{7}{8}$ m tall.

Which tree seedling is taller?

Exercise 4 • page 43

9-4 Comparing Fractions with Like Numerators

1 What fraction of each shape is colored?
What fraction is not colored?

(a)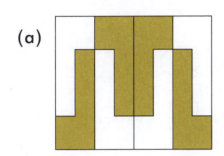

(b)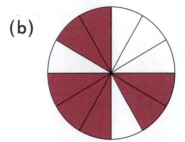

2 (a) $\frac{5}{7}$ and $\frac{}{}$ make 1.

(b) $\frac{5}{8}$ is $\frac{}{}$ more than $\frac{3}{8}$.

(c) $\frac{2}{10}$ is $\frac{}{}$ less than $\frac{7}{10}$.

(d) $\frac{6}{} = 1$

3 What fraction is indicated by each letter on the number line?

(a)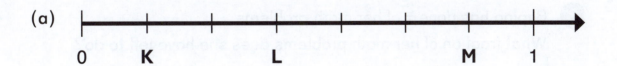

(b)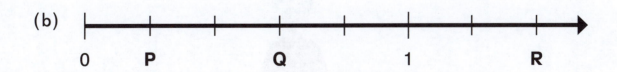

4 Which of these fractions are greater than 1?

$\frac{5}{7}$ $\frac{5}{3}$ $\frac{6}{6}$ $\frac{7}{8}$ $\frac{9}{7}$

5 What sign, >, <, or =, goes in the ◯?

(a) $\frac{3}{7}$ ◯ $\frac{3}{4}$ (b) $\frac{5}{8}$ ◯ $\frac{3}{8}$

(c) $\frac{6}{6}$ ◯ $\frac{1}{3}$ (d) $\frac{5}{3}$ ◯ $\frac{2}{3}$

(e) $\frac{5}{5}$ ◯ $\frac{10}{10}$ (f) $\frac{7}{4}$ ◯ $\frac{4}{7}$

6 Put the numbers in order from least to greatest.

(a) $\frac{2}{7}, \frac{5}{7}, \frac{7}{3}$ (b) $\frac{5}{7}, \frac{5}{9}, \frac{5}{12}$

(c) $\frac{5}{3}, 1, \frac{1}{3}, 0$ (d) $\frac{7}{5}, \frac{7}{8}, \frac{4}{8}, \frac{4}{9}$

7 What fraction has a numerator of 3, is greater than $\frac{3}{5}$, and is less than 1?

8 Papina has done $\frac{2}{3}$ of her math problems.
What fraction of her math problems does she have left to do?

Exercise 5 • page 47

9-5 Practice

Chapter 10

Fractions — Part 2

Think

Use three strips of paper that are all equal in length and width.
Fold all three strips in half, open them up again, draw a line at the crease, and color half of each.

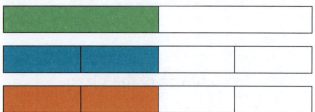

The first fold divides the strip of paper into 2 equal parts.

Fold the second and third strips in half again.

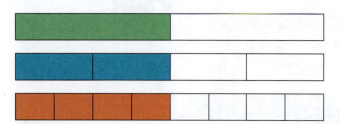

Fold the third strip in half again.

Compare the three strips.
How many equal parts does each strip have?
How many parts are colored on each strip?
What fraction of each strip is colored?
What can you say about the three fractions?

Learn

1 out of 2 parts are colored.

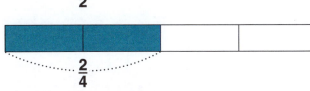

2 out of 4 parts are colored.

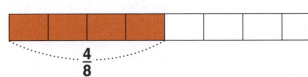

4 out of 8 parts are colored.

Each strip has the same amount colored.
These fractions are equal even though the numerators
and denominators are different.

$$\frac{1}{2} = \frac{2}{4} = \frac{4}{8}$$

$\frac{1}{2}$, $\frac{2}{4}$, and $\frac{4}{8}$ are **equivalent fractions**.

Equivalent fractions have the same value.

We can use number lines to show
equivalent fractions.

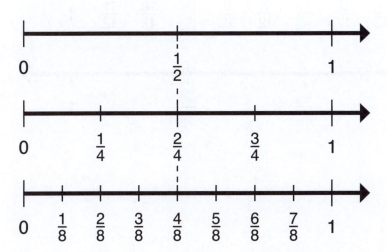

Do you notice any other equivalent fractions?

Do

1

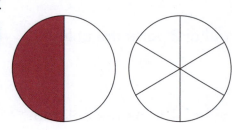

What fraction with a denominator of 6 is equivalent to $\frac{1}{2}$?

2 $\frac{2}{3}$ of the bar is colored.

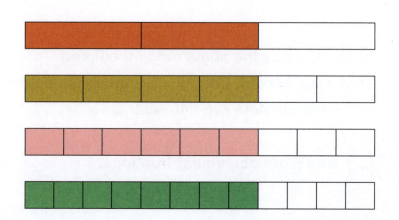

(a) $\frac{2}{3} = \frac{\blacksquare}{6}$

(b) $\frac{2}{3} = \frac{\blacksquare}{9}$

(c) $\frac{2}{3} = \frac{\blacksquare}{12}$

3 (a) What are the missing numbers on the number lines?

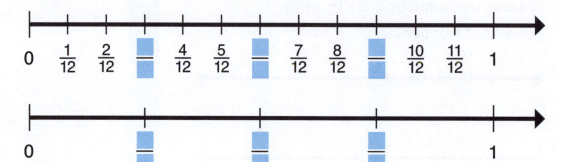

(b) $\frac{3}{12} = \frac{1}{\blacksquare}$

(c) $\frac{9}{12} = \frac{3}{\blacksquare}$

(d) $\frac{1}{2} = \frac{\blacksquare}{12}$

4 (a) $\dfrac{5}{10} = \dfrac{\boxed{}}{2}$

(b) $\dfrac{4}{10} = \dfrac{2}{\boxed{}}$

5 Use the fraction chart to find pairs of equivalent fractions.

1				

$\frac{1}{2}$	$\frac{1}{2}$

$\frac{1}{3}$	$\frac{1}{3}$	$\frac{1}{3}$

$\frac{1}{4}$	$\frac{1}{4}$	$\frac{1}{4}$	$\frac{1}{4}$

$\frac{1}{5}$	$\frac{1}{5}$	$\frac{1}{5}$	$\frac{1}{5}$	$\frac{1}{5}$

$\frac{1}{6}$	$\frac{1}{6}$	$\frac{1}{6}$	$\frac{1}{6}$	$\frac{1}{6}$	$\frac{1}{6}$

$\frac{1}{7}$	$\frac{1}{7}$	$\frac{1}{7}$	$\frac{1}{7}$	$\frac{1}{7}$	$\frac{1}{7}$	$\frac{1}{7}$

$\frac{1}{8}$	$\frac{1}{8}$	$\frac{1}{8}$	$\frac{1}{8}$	$\frac{1}{8}$	$\frac{1}{8}$	$\frac{1}{8}$	$\frac{1}{8}$

$\frac{1}{9}$	$\frac{1}{9}$	$\frac{1}{9}$	$\frac{1}{9}$	$\frac{1}{9}$	$\frac{1}{9}$	$\frac{1}{9}$	$\frac{1}{9}$	$\frac{1}{9}$

$\frac{1}{10}$	$\frac{1}{10}$	$\frac{1}{10}$	$\frac{1}{10}$	$\frac{1}{10}$	$\frac{1}{10}$	$\frac{1}{10}$	$\frac{1}{10}$	$\frac{1}{10}$	$\frac{1}{10}$

Exercise 1 • page 51

Think

Compare the bars and the numerators and denominators of the fractions in these two sets of equivalent fractions.
What do you notice?

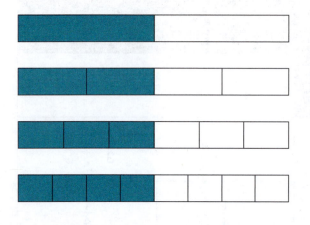

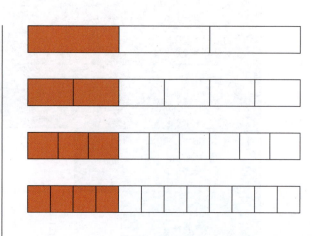

$$\frac{1}{2} = \frac{2}{4} = \frac{3}{6} = \frac{4}{8}$$

$$\frac{1}{3} = \frac{2}{6} = \frac{3}{9} = \frac{4}{12}$$

Learn

When the number of shaded parts doubled, so did the total number of parts.

The number of blue and orange parts are the same in each row, but the number of total parts are not the same.

Both the number of shaded parts and the number of total parts are 2, 3, and 4 times as many, but the size of the shaded part does not change...

I noticed that

$$\overset{\times\,4}{\frac{1}{2} = \frac{4}{8}}\underset{\times\,4}{} \text{ and } \overset{\times\,4}{\frac{1}{3} = \frac{4}{12}}\underset{\times\,4}{}.$$

To find equivalent fractions, we can multiply both the numerator and denominator by the same number.

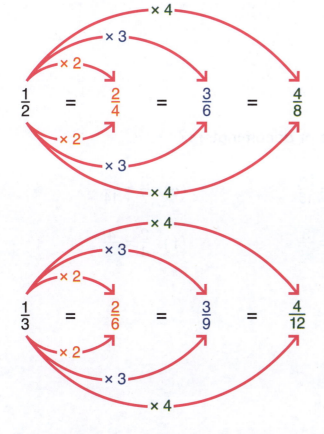

Find other equivalent fractions for $\frac{1}{2}$ and $\frac{1}{3}$.

Do

1 What is the missing numerator and denominator?

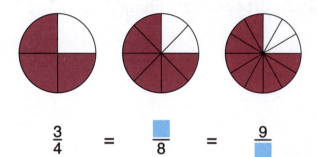

$$\frac{3}{4} \quad = \quad \frac{\square}{8} \quad = \quad \frac{9}{\square}$$

2 What are the missing numbers?

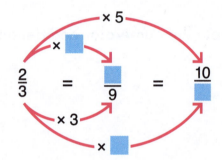

$$\frac{2}{3} \quad = \quad \frac{\square}{9} \quad = \quad \frac{10}{\square}$$

3 What are the missing numerators or denominators?

(a) $\frac{1}{4} = \frac{\square}{12}$

(b) $\frac{2}{5} = \frac{\square}{15}$

(c) $\frac{3}{7} = \frac{\square}{14}$

(d) $\frac{2}{6} = \frac{6}{\square}$

(e) $\frac{1}{8} = \frac{3}{\square}$

(f) $\frac{2}{4} = \frac{4}{\square}$

4 List some equivalent fractions.

(a) $\frac{1}{5}$ → $\frac{2}{10},\ \frac{\square}{\square},\ \frac{\square}{\square},\ \frac{\square}{\square}$

(b) $\frac{2}{7}$ → $\frac{\square}{\square},\ \frac{\square}{\square},\ \frac{\square}{\square},\ \frac{\square}{\square}$

Exercise 2 • page 55

Think

A paper strip is divided into 12 parts and $\frac{8}{12}$ of it is colored.

Can we color the same amount on another paper strip of the same length and width but divide it into fewer parts first?

How can we find an equivalent fraction of $\frac{8}{12}$ with a denominator less than 12?

Learn

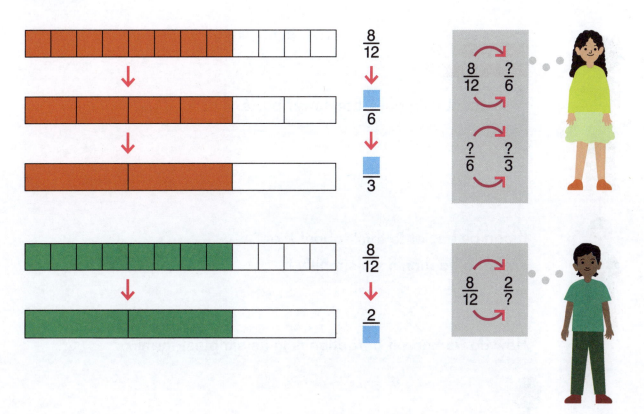

We can divide both the numerator and denominator by the same number to get equivalent fractions.

When we do so, we are **simplifying** the fraction.

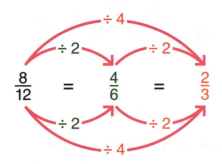

$\frac{2}{3}$ is the simplest equivalent fraction for both $\frac{8}{12}$ and $\frac{4}{6}$.

When a fraction cannot be simplified any further, the fraction is in its **simplest form**.

The simplest form of $\frac{8}{12}$ is 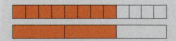.

$\frac{8}{12}$ is equal to $\frac{2}{3}$, but $\frac{2}{3}$ has fewer parts.

It can be easier to think about the size of a fraction if we simplify it.

How do we know if a fraction is in its simplest form?

Do

1 (a) What is the missing numerator and denominator?

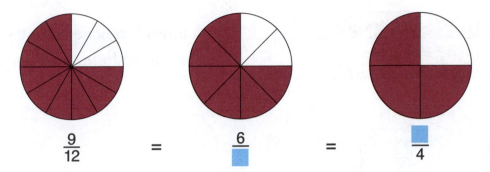

$$\frac{9}{12} \qquad = \qquad \frac{6}{\,\blacksquare\,} \qquad = \qquad \frac{\blacksquare}{4}$$

(b) The simplest form of $\frac{9}{12}$ is $\frac{\blacksquare}{\blacksquare}$.

2 The line is $\frac{8}{16}$ inches long.

What other ways can we write the length of the line?

$$\frac{8}{16} \text{ in} = \frac{4}{\blacksquare} \text{ in} = \frac{2}{\blacksquare} \text{ in} = \frac{1}{\blacksquare} \text{ in}$$

3 Find the missing numerators or denominators.

(a) $\dfrac{8}{10} = \dfrac{\blacksquare}{5}$

(b) $\dfrac{3}{9} = \dfrac{\blacksquare}{3}$

(c) $\dfrac{10}{12} = \dfrac{\blacksquare}{6}$

(d) $\dfrac{6}{12} = \dfrac{3}{\blacksquare}$

(e) $\dfrac{6}{8} = \dfrac{3}{\blacksquare}$

(f) $\dfrac{5}{10} = \dfrac{1}{\blacksquare}$

4 How can we use multiplication or division to show that $\frac{8}{12}$ and $\frac{6}{9}$ are equivalent fractions?

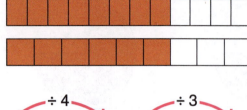
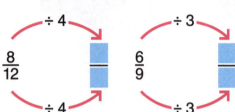

If two fractions have the same simplest form, they are equal.

$\div 4$ $\div 3$

$\frac{8}{12}$ $\frac{6}{9}$

$\div 4$ $\div 3$

5 Express each fraction in simplest form.

(a) $\frac{4}{6}$

(b) $\frac{5}{10}$

(c) $\frac{6}{8}$

(d) $\frac{6}{12}$

(e) $\frac{3}{15}$

(f) $\frac{10}{12}$

6 Which of the following fractions have $\frac{1}{2}$ as simplest form?

$\frac{8}{10}$ $\frac{5}{10}$ $\frac{4}{9}$ $\frac{7}{12}$ $\frac{4}{8}$ $\frac{7}{14}$ $\frac{3}{7}$ $\frac{3}{6}$

How can you easily tell if a fraction is equal to $\frac{1}{2}$?

Exercise 3 • page 58

Think

The friends are wrapping presents.

Mei's ribbon is $\frac{3}{4}$ m long.

Dion's ribbon is $\frac{6}{7}$ m long.

Emma's ribbon is $\frac{5}{8}$ m long.

$\frac{3}{4}$	$\frac{6}{7}$	$\frac{5}{8}$
Mei	Dion	Emma

How can we put these fractions in order from least to greatest?
Whose ribbon is the shortest?
Whose ribbon is the longest?

How can you use the numerators and denominators to help you put them in order?

Learn

$\frac{3}{4} = \frac{6}{8}$

$\frac{5}{8} < \frac{6}{8}$, so $\frac{5}{8} < \frac{3}{4}$.

$\frac{5}{8}$	$\frac{3}{4}$
Emma	Mei

I compared Mei's ribbon and Emma's ribbon by using equivalent fractions to make the denominators the same.

Emma's ribbon is shorter than Mei's ribbon.

$\frac{3}{4} = \frac{6}{8}$

$\frac{6}{8} < \frac{6}{7}$, so $\frac{3}{4} < \frac{6}{7}$.

$\frac{3}{4}$	$\frac{6}{7}$
Mei	Dion

I compared Mei's ribbon and Dion's ribbon by using equivalent fractions to make the numerators the same.

Mei's ribbon is shorter than Dion's ribbon.

_____'s ribbon is the shortest.

_____'s ribbon is the longest.

$\frac{5}{8}$	$\frac{3}{4}$	$\frac{6}{7}$
Emma	Mei	Dion

We can compare fractions that do not have the same numerator or denominator by using equivalent fractions.

Emma	$\frac{1}{8}$	$\frac{1}{8}$	$\frac{1}{8}$	$\frac{1}{8}$	$\frac{1}{8}$	$\frac{1}{8}$	$\frac{1}{8}$	$\frac{1}{8}$
Mei	$\frac{1}{4}$		$\frac{1}{4}$		$\frac{1}{4}$		$\frac{1}{4}$	
Dion	$\frac{1}{7}$	$\frac{1}{7}$	$\frac{1}{7}$	$\frac{1}{7}$	$\frac{1}{7}$	$\frac{1}{7}$	$\frac{1}{7}$	

Do

1 Which is greater, $\frac{3}{5}$ or $\frac{7}{10}$?

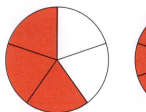 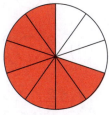

$$\frac{3}{5} = \frac{?}{10}$$

2 Which is less, $\frac{2}{3}$ or $\frac{5}{9}$?

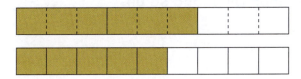

$$\frac{2}{3} = \frac{?}{9}$$

3 What sign, > or <, goes in the ◯?

(a) $\frac{2}{3}$ ◯ $\frac{5}{6}$

(b) $\frac{1}{2}$ ◯ $\frac{3}{8}$

(c) $\frac{2}{9}$ ◯ $\frac{1}{3}$

(d) $\frac{3}{4}$ ◯ $\frac{5}{8}$

(e) $\frac{2}{5}$ ◯ $\frac{3}{10}$

(f) $\frac{5}{12}$ ◯ $\frac{1}{4}$

4 Which is less, $\frac{2}{5}$ or $\frac{4}{9}$?

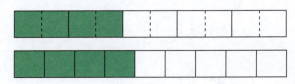

$$\frac{2}{5} = \frac{4}{?}$$

5 What sign, >, <, or =, goes in the ◯?

(a) $\frac{1}{2}$ ◯ $\frac{3}{6}$

(b) $\frac{6}{7}$ ◯ $\frac{3}{5}$

(c) $\frac{6}{9}$ ◯ $\frac{2}{7}$

(d) $\frac{3}{7}$ ◯ $\frac{9}{8}$

(e) $\frac{10}{11}$ ◯ $\frac{5}{8}$

(f) $\frac{2}{3}$ ◯ $\frac{6}{9}$

6 (a) List equivalent fractions for $\frac{2}{3}$ and $\frac{3}{4}$ by multiplying the numerator and denominator first by 2, then by 3, and so on.

$\frac{2}{3}$ → $\frac{2}{3}, \frac{2}{6}, \frac{2}{9}, \frac{2}{12}$

$\frac{3}{4}$ → $\frac{3}{4}, \frac{3}{8}, \frac{3}{12}, \frac{3}{16}$

(b) Which is greater, $\frac{2}{3}$ or $\frac{3}{4}$?

$\frac{8}{12} < \frac{9}{12}$, so...

$\frac{6}{9} < \frac{6}{8}$, so...

7 What sign, > or <, goes in the ◯?

(a) $\frac{2}{3}$ ◯ $\frac{3}{5}$

(b) $\frac{1}{2}$ ◯ $\frac{2}{3}$

(c) $\frac{5}{6}$ ◯ $\frac{3}{8}$

(d) $\frac{3}{4}$ ◯ $\frac{4}{6}$

Exercise 4 • page 62

Think

Alex skateboarded $\frac{5}{6}$ mile.

Sofia skateboarded $\frac{2}{5}$ mile.

Mei skateboarded $\frac{8}{9}$ mile.

$\frac{5}{6}$	$\frac{2}{5}$	$\frac{8}{9}$
Alex	Sofia	Mei

How can we put these fractions in order from least to greatest?
Who skateboarded the longest distance?
Who skateboarded the shortest distance?

Which fractions are greater than $\frac{1}{2}$ and which are less than $\frac{1}{2}$?

Which fractions are closer to 1?

Learn

$\frac{2}{5} < \frac{1}{2}$ since $\frac{2}{5} < \frac{2}{4}$.

$\frac{5}{6} > \frac{1}{2}$ since $\frac{5}{6} > \frac{3}{6}$.

$\frac{8}{9} > \frac{1}{2}$ because it is almost 1.

| $\frac{2}{5}$ |
| Sofia |

$\frac{2}{5}$ is the only one of the three fractions that is less than $\frac{1}{2}$.

$\frac{5}{6}$ and $\frac{1}{6}$ make 1.

$\frac{8}{9}$ and $\frac{1}{9}$ make 1.

$\frac{1}{9} < \frac{1}{6}$, so $\frac{5}{6} < \frac{8}{9}$.

| $\frac{5}{6}$ | $\frac{8}{9}$ |
| Alex | Mei |

I compared $\frac{5}{6}$ and $\frac{8}{9}$ to 1. $\frac{8}{9}$ is closer to 1 than $\frac{5}{6}$.

_____ skateboarded the longest distance.

_____ skateboarded the shortest distance.

| $\frac{2}{5}$ | $\frac{5}{6}$ | $\frac{8}{9}$ |
| Sofia | Alex | Mei |

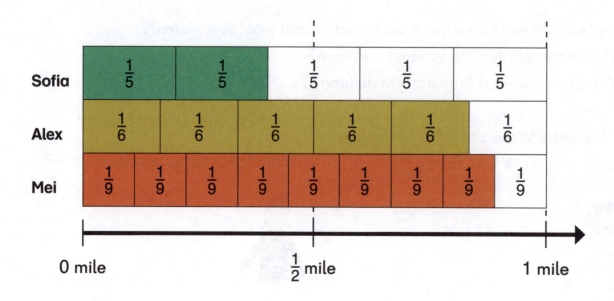

Do

1 Which is greater, $\frac{1}{2}$ or $\frac{5}{8}$?

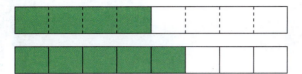

$\frac{4}{8} = \frac{1}{2}$, and $\frac{5}{8} > \frac{4}{8}$, so...

2 Which is greater, $\frac{1}{2}$ or $\frac{4}{9}$?

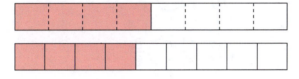

$\frac{4}{8} = \frac{1}{2}$, and $\frac{4}{9} < \frac{4}{8}$, so...

3 Which is greater, $\frac{5}{8}$ or $\frac{4}{9}$?

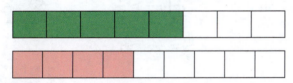

$\frac{4}{9} < \frac{1}{2}$, and $\frac{1}{2} < \frac{5}{8}$, so...

4 Which of the following are less than $\frac{1}{2}$?

$\frac{5}{6}$ $\frac{3}{8}$ $\frac{7}{12}$ $\frac{3}{10}$ $\frac{5}{11}$ $\frac{2}{3}$ $\frac{2}{5}$ $\frac{3}{7}$

5 What sign, > or <, goes in the $\bigcirc$?

(a) $\frac{1}{4} \bigcirc \frac{5}{7}$ (b) $\frac{2}{3} \bigcirc \frac{1}{5}$ (c) $\frac{1}{3} \bigcirc \frac{5}{8}$

6 Which is less, $\frac{6}{7}$ or $\frac{10}{11}$?

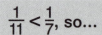

$\frac{1}{11} < \frac{1}{7}$, so...

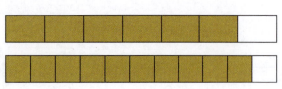

7 What sign, > or <, goes in the ◯?

(a) $\frac{9}{10}$ ◯ $\frac{5}{6}$

(b) $\frac{7}{8}$ ◯ $\frac{2}{3}$

(c) $\frac{3}{4}$ ◯ $\frac{5}{6}$

(d) $\frac{11}{12}$ ◯ $\frac{8}{9}$

8 Put the fractions $\frac{2}{3}$, $\frac{1}{4}$, and $\frac{5}{12}$ in order from least to greatest.

I used equivalent fractions: $\frac{8}{12}$, $\frac{3}{12}$, and $\frac{5}{12}$, so...

Only $\frac{2}{3} > \frac{1}{2}$.
$\frac{1}{4} = \frac{3}{12}$, so...

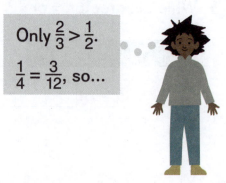

9 Put the fractions in order from least to greatest.

(a) $\frac{5}{8}$, $\frac{1}{2}$, $\frac{3}{4}$

(b) $\frac{7}{8}$, $\frac{2}{5}$, $\frac{7}{10}$

(c) $\frac{1}{2}$, $\frac{3}{8}$, $\frac{3}{4}$, $\frac{5}{6}$

(d) $\frac{5}{6}$, $\frac{1}{3}$, $\frac{1}{2}$, $\frac{7}{8}$

Exercise 5 • page 65

1 Find the missing numerators.

(a) $\frac{1}{5} = \frac{\square}{10}$

(b) $\frac{6}{9} = \frac{\square}{3}$

(c) $\frac{1}{2} = \frac{\square}{4} = \frac{\square}{6}$

(d) $\frac{9}{15} = \frac{\square}{5}$

(e) $\frac{\square}{4} = \frac{6}{8}$

(f) $\frac{1}{3} = \frac{\square}{6} = \frac{\square}{9}$

2 Find the missing denominators.

(a) $\frac{3}{5} = \frac{6}{\square}$

(b) $\frac{2}{3} = \frac{10}{\square}$

(c) $\frac{1}{2} = \frac{4}{\square} = \frac{6}{\square}$

(d) $\frac{3}{\square} = \frac{9}{12}$

(e) $\frac{2}{\square} = \frac{1}{7}$

(f) $\frac{2}{3} = \frac{4}{\square} = \frac{6}{\square}$

3 (a) Write two fractions that are equivalent to $\frac{4}{6}$.

(b) Write two fractions that are equivalent to $\frac{1}{2}$.

(c) Write two fractions that are equivalent to 1.

4 What number is indicated by each letter on the number line?
Give your answers in simplest form.

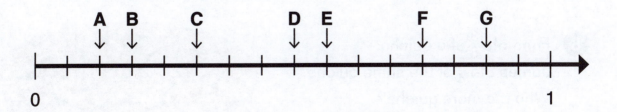

5 Write each of the following fractions in simplest form.
Some of them are already in simplest form.

(a) $\frac{6}{10}$

(b) $\frac{4}{8}$

(c) $\frac{5}{10}$

(d) $\frac{6}{9}$

(e) $\frac{10}{12}$

(f) $\frac{7}{8}$

(g) $\frac{9}{10}$

(h) $\frac{3}{4}$

(i) $\frac{10}{15}$

(j) $\frac{7}{9}$

(k) $\frac{8}{16}$

(l) $\frac{3}{6}$

6 What sign, > or <, goes in the $\bigcirc$?

(a) $\frac{1}{3} \bigcirc \frac{5}{6}$

(b) $\frac{5}{8} \bigcirc \frac{11}{16}$

(c) $\frac{2}{7} \bigcirc \frac{4}{3}$

(d) $\frac{3}{5} \bigcirc \frac{1}{3}$

(e) $\frac{2}{3} \bigcirc \frac{3}{2}$

(f) $\frac{9}{10} \bigcirc \frac{7}{8}$

7 Put the fractions in order from least to greatest.

(a) $\frac{2}{3}, \frac{3}{4}, \frac{7}{12}$

(b) $\frac{5}{7}, \frac{5}{9}, \frac{1}{2}$

(c) $\frac{7}{5}, \frac{7}{8}, \frac{4}{8}, \frac{7}{9}$

(d) $\frac{1}{2}, \frac{3}{10}, \frac{6}{5}, \frac{3}{8}$

8 Fang ate $\frac{3}{8}$ of a quiche.
James ate $\frac{1}{4}$ of the same quiche.
Who ate more quiche?

Exercise 6 • page 68

Lesson 7
Adding and Subtracting Fractions — Part 1

Think

Sofia has $\frac{5}{8}$ gallons of red paint.

Emma has $\frac{2}{8}$ gallons of blue paint.

(a) How much paint do they have altogether?

(b) Sofia used $\frac{2}{8}$ gallons to paint a door.
How much paint does she have left?

Learn

(a)

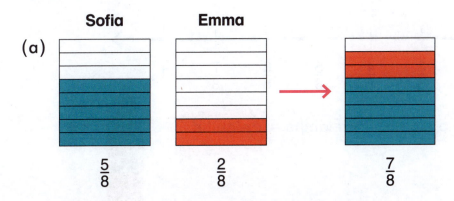

How many one-eighths make $\frac{5}{8}$ and $\frac{2}{8}$?

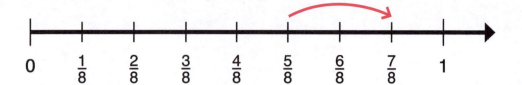

$\frac{5}{8} + \frac{2}{8} = \frac{\boxed{}}{8}$

5 eighths + 2 eighths = ? eighths

They have $\frac{\boxed{}}{\boxed{}}$ gallons of paint altogether.

(b)

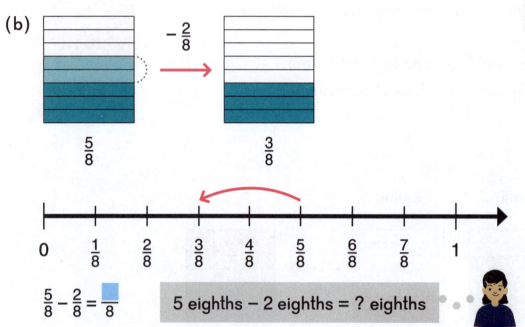

$\frac{5}{8} - \frac{2}{8} = \frac{\boxed{}}{8}$

5 eighths − 2 eighths = ? eighths

Sofia has $\frac{\boxed{}}{\boxed{}}$ gallons of paint left.

When adding fractions with the same denominator, add the numerators.
When subtracting fractions with the same denominator, subtract the numerators.

We can do this because we are adding or subtracting equal units.

Do

1 Add $\frac{3}{7}$ and $\frac{2}{7}$.

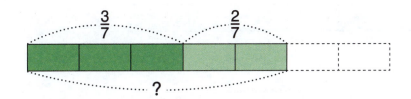

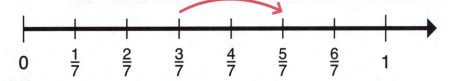

$\frac{3}{7} + \frac{2}{7} = \frac{\square}{7}$

3 sevenths + 2 sevenths = ? sevenths

2 Subtract $\frac{3}{9}$ from $\frac{7}{9}$.

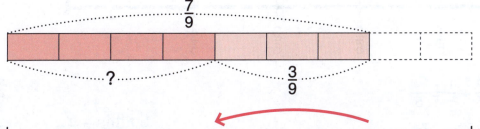

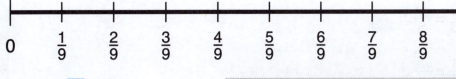

$\frac{7}{9} - \frac{3}{9} = \frac{\square}{9}$

7 ninths − 3 ninths = ? ninths

3 Subtract $\frac{3}{4}$ from 1.

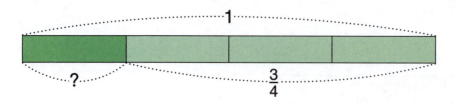

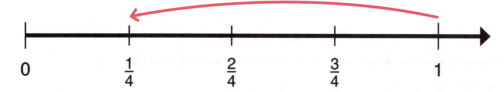

$1 - \frac{3}{4} = \frac{\square}{4}$

1 = ? fourths

4

5 fifths = ?

(a) $\frac{2}{5} + \frac{1}{5} = \frac{\square}{5}$

(b) $\frac{3}{5} + \frac{2}{5} = \frac{\square}{5} = \square$

(c) $\frac{4}{5} - \frac{1}{5} = \frac{\square}{5}$

(d) $\frac{5}{5} - \frac{1}{5} = \frac{\square}{5}$

If the numerator is 0, the fraction is equal to 0.

(e) $\frac{2}{5} - \frac{2}{5} = \frac{\square}{5} = \square$

(f) $1 - \frac{2}{5} = \frac{\square}{5}$

5 Find the value.

(a) $\frac{2}{4} + \frac{1}{4}$

(b) $\frac{3}{8} + \frac{4}{8}$

(c) $\frac{5}{6} - \frac{4}{6}$

(d) $\frac{8}{10} - \frac{5}{10}$

(e) $1 - \frac{3}{10}$

(f) $\frac{1}{12} + \frac{5}{12} + \frac{5}{12}$

6 Find the missing numbers.

(a) $\frac{3}{9} + \frac{\boxed{}}{9} = \frac{5}{9}$

(b) $\frac{\boxed{}}{7} + \frac{3}{7} = \frac{6}{7}$

(c) $\frac{3}{4} - \frac{\boxed{}}{4} = \frac{1}{4}$

(d) $\frac{\boxed{}}{12} - \frac{6}{12} = \frac{5}{12}$

7 Papina mowed $\frac{1}{3}$ of the lawn so far.
How much more of the lawn does she have left to mow?

8 Phyllis bought $\frac{3}{10}$ kg of green grapes, $\frac{4}{10}$ kg of red grapes, and $\frac{2}{10}$ kg of purple grapes.
How many kilograms of grapes did she buy in all?

Exercise 7 • page 71

Think

Dion had $\frac{8}{9}$ gallons of paint.

He used $\frac{2}{9}$ gallons to paint a chair.

How much paint does he have left?
Give your answer in simplest form.

Learn

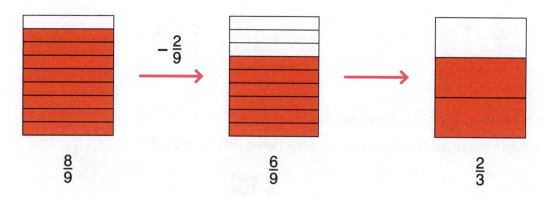

$$\frac{8}{9} \qquad \frac{6}{9} \qquad \frac{2}{3}$$

$$\frac{8}{9} - \frac{2}{9} = \frac{6}{9} = \frac{2}{3}$$

He has $\frac{}{}$ gallons of paint left.

If we give the final answer in simplest form, it is easier to see how much paint is left. Put final answers in simplest form.

<u>Do</u>

1 Add $\frac{1}{6}$ and $\frac{3}{6}$.

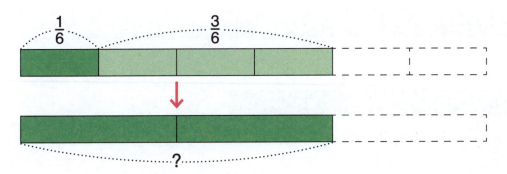

$$\frac{1}{6} + \frac{3}{6} = \frac{\square}{6} = \frac{\square}{\square}$$

2 Subtract $\frac{1}{4}$ from $\frac{3}{4}$.

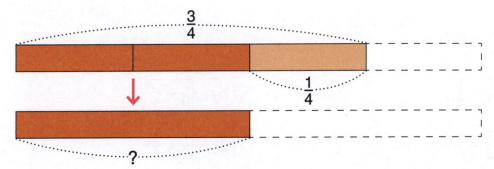

$$\frac{3}{4} - \frac{1}{4} = \frac{\square}{4} = \frac{\square}{\square}$$

3 (a) $\frac{3}{8} + \frac{1}{8} = \frac{\square}{8} = \frac{\square}{\square}$ (b) $\frac{3}{10} + \frac{5}{10} = \frac{\square}{10} = \frac{\square}{\square}$

(c) $\frac{5}{6} - \frac{1}{6} = \frac{\square}{6} = \frac{\square}{\square}$ (d) $\frac{11}{12} - \frac{7}{12} = \frac{\square}{12} = \frac{\square}{\square}$

4 Subtract $\frac{2}{6}$ from 1.

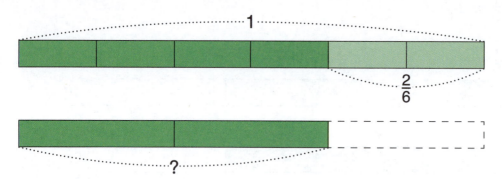

$1 - \frac{2}{6} = \frac{\blacksquare}{6} = \frac{\blacksquare}{\blacksquare}$

5 (a) $1 - \frac{3}{9} = \frac{\blacksquare}{9} = \frac{\blacksquare}{\blacksquare}$

(b) $1 - \frac{2}{12} = \frac{\blacksquare}{12} = \frac{\blacksquare}{\blacksquare}$

6 Find the value.

(a) $\frac{5}{8} + \frac{1}{8}$

(b) $\frac{1}{10} + \frac{7}{10}$

(c) $\frac{7}{9} - \frac{4}{9}$

(d) $\frac{11}{15} - \frac{6}{15}$

7 Randy bought $\frac{7}{16}$ lb of yellow cheese and $\frac{3}{16}$ lb of white cheese.

(a) How much cheese did he buy in all?

(b) How much more yellow cheese did he buy than white cheese?

Exercise 8 • page 74

1 Find the value.

(a) $\frac{3}{8} + \frac{2}{8}$

(b) $\frac{2}{7} + \frac{5}{7}$

(c) $\frac{4}{9} + \frac{5}{9}$

(d) $\frac{2}{6} + \frac{3}{6}$

(e) $\frac{3}{10} + \frac{3}{10}$

(f) $\frac{4}{15} + \frac{6}{15}$

(g) $\frac{1}{6} + \frac{1}{6} + \frac{1}{6}$

(h) $\frac{1}{10} + \frac{7}{10} + \frac{2}{10}$

(i) $\frac{1}{16} + \frac{7}{16} + \frac{2}{16}$

(j) $\frac{7}{9} - \frac{3}{9}$

(k) $\frac{5}{6} - \frac{2}{6}$

(l) $\frac{4}{5} - \frac{1}{5}$

(m) $\frac{9}{10} - \frac{3}{10}$

(n) $\frac{5}{12} - \frac{2}{12}$

(o) $\frac{8}{15} - \frac{3}{15}$

(p) $1 - \frac{2}{6}$

(q) $1 - \frac{5}{8}$

(r) $\frac{13}{16} - \frac{5}{16}$

2 What sign, >, <, or =, goes in the ◯?

(a) $\frac{1}{3} + \frac{2}{3} \bigcirc \frac{5}{6} + \frac{1}{6}$

(b) $\frac{9}{10} - \frac{6}{10} \bigcirc 1 - \frac{3}{9}$

(c) $1 - \frac{2}{7} \bigcirc \frac{2}{8} + \frac{3}{8}$

(d) $\frac{3}{8} + \frac{2}{8} + \frac{1}{8} \bigcirc \frac{3}{2}$

(e) $\frac{5}{8} - \frac{3}{8} \bigcirc \frac{11}{16} - \frac{7}{16}$

(f) $\frac{7}{9} - \frac{1}{9} \bigcirc \frac{1}{12} + \frac{1}{12} + \frac{1}{12} + \frac{1}{12}$

3 Kalama ran $\frac{7}{8}$ mile and Hailey ran $\frac{5}{8}$ mile.
Who ran farther and by how much?

4 Nicole bought $\frac{5}{12}$ ft of blue lace, $\frac{3}{12}$ ft of green lace, and $\frac{2}{12}$ ft of yellow lace.
How many feet of lace does she have in all?

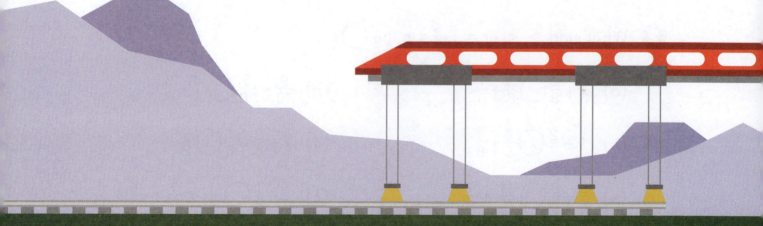

5 Pekelo had 1 m of ribbon.
He used $\frac{3}{8}$ m to wrap a present and $\frac{1}{8}$ m to make a bow.
How much ribbon does he have left?

6 A track-laying machine laid $\frac{8}{9}$ km of track on Monday and $\frac{5}{9}$ km of track on Tuesday. How many more kilometers of track did it lay on Monday than on Tuesday?

Exercise 9 • page 77

10-9 Practice B

Chapter 11

Measurement

Think

(a) What is Alex's height in meters and centimeters?

(b) What is Emma's height in centimeters?

Learn

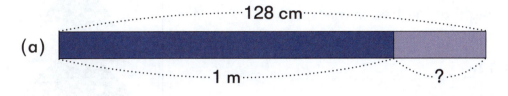

(a)

128 cm = 100 cm + 28 cm
        = 1 m 28 cm

1 m = 100 cm

Alex's height is ▢ m ▢ cm.

(b)

1 m 30 cm = 100 cm + 30 cm
          = 130 cm

Emma's height is ▢ cm.

Emma is 2 cm taller than Alex.

When we express a length in two units, such as 1 m 30 cm, we are expressing the length in **compound units**.

Do

1 Write the height of the tree in meters.

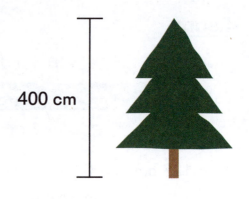

400 cm

100 cm = 1 m

400 cm = [ ] m

2 Write the height of the giraffe in centimeters.

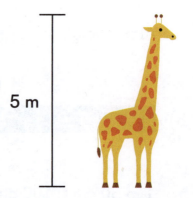

5 m

1 m = 100 cm

5 m = [ ] cm

3 A desk is 165 cm long.
Write its length in meters and centimeters.

165 cm = 100 cm + [ ] cm

= [ ] m [ ] cm

165 cm
/   \
100 cm   65 cm

4 A string is 250 cm long.
Write its length in meters and centimeters.

250 cm = 200 cm + [ ] cm

= [ ] m [ ] cm

100 cm = 1 m
200 cm = ? m

5 A whiteboard is 1 m 20 cm high and 3 m 45 cm long.
Write its height and length in centimeters.

1 m 20 cm = 100 cm + ▭ cm

= ▭ cm

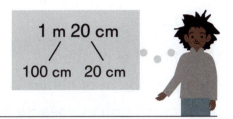

1 m 20 cm
/   \
100 cm   20 cm

3 m 45 cm = ▭ cm + 45 cm

= ▭ cm

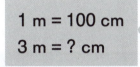

1 m = 100 cm
3 m = ? cm

6 Write in meters and centimeters.

(a) 320 cm

(b) 443 cm

(c) 982 cm

7 Write in centimeters.

(a) 2 m 70 cm

(b) 5 m 86 cm

(c) 7 m 5 cm

8 The table shows the heights of some animals.
Put the heights in order from shortest to tallest.

Animal	Height
Antelope	180 cm
Gazelle	1 m 65 cm
Moose	210 cm
Zebra	1 m 50 cm

Exercise 1 • page 81

Think

Alex has 3 m of ribbon.

He uses 75 cm of ribbon to wrap a present.

How much ribbon does he have left?

Express your answer in compound units.

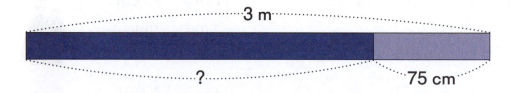

1 m = 100 cm
I will use mental math for making 100.

3 m − 75 cm = ?

What different methods are there to find the answer?
I wonder which method is the easiest.

Learn

$3\ m - 75\ cm = ?$

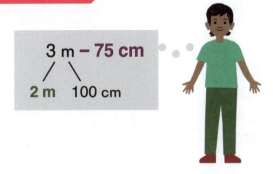

$$3\ m - 75\ cm$$
$$2\ m \quad 100\ cm$$

$$100\ cm \xrightarrow{-75\ cm} 25\ cm \xrightarrow{+2\ m} 2\ m\ 25\ cm$$

Method 2

$1\ m = 75\ cm + 25\ cm$
1 m is **25 cm** more than 75 cm.

$$3\ m \xrightarrow{-1\ m} 2\ m \xrightarrow{+25\ cm} 2\ m\ 25\ cm$$

Method 3

$300\ cm - 75\ cm = 225\ cm$
$225\ cm = 2\ m\ 25\ cm$

$3\ m = 300\ cm$
$300 - 75 = ?$

$3\ m - 75\ cm =$ ☐ m ☐ cm

Alex has ☐ m ☐ cm of ribbon left.

Do

1 Find pairs of lengths that make 1 m.

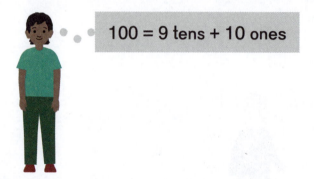

38 cm	45 cm	55 cm	71 cm
76 cm	29 cm	62 cm	58 cm
34 cm	42 cm	24 cm	66 cm

100 = 9 tens + 10 ones

2 (a) Subtract 35 cm from 1 m.

1 m − 35 cm = [ ] cm

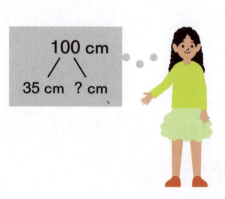

100 cm
/ \
35 cm   ? cm

(b) Subtract 35 cm from 3 m.

3 m − 35 cm = [ ] m [ ] cm

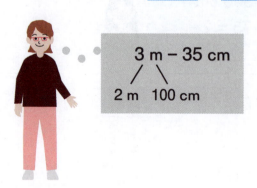

3 m − 35 cm
/ \
2 m   100 cm

(c) Subtract 1 m 35 cm from 4 m.

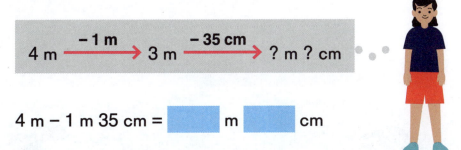

4 m – 1 m 35 cm = ▭ m ▭ cm

(d) Subtract 3 m 95 cm from 7 m.

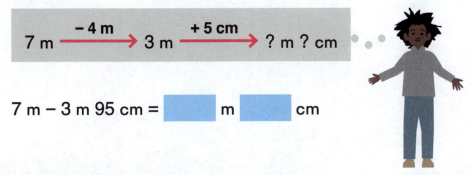

7 m – 3 m 95 cm = ▭ m ▭ cm

3 Subtract in compound units.

(a) 1 m – 36 cm

(b) 2 m – 70 cm

(c) 4 m – 87 cm

(d) 3 m – 1 m 25 cm

(e) 10 m – 6 m 42 cm

(f) 30 m – 25 m 50 cm

4 (a) 1 m = 55 cm + ▭ cm

(b) 2 m 29 cm + ▭ cm = 3 m

(c) 2 m = 1 m 58 cm + ▭ cm

(d) 5 m = 3 m 75 cm + 1 m ▭ cm

Exercise 2 · page 85

Think

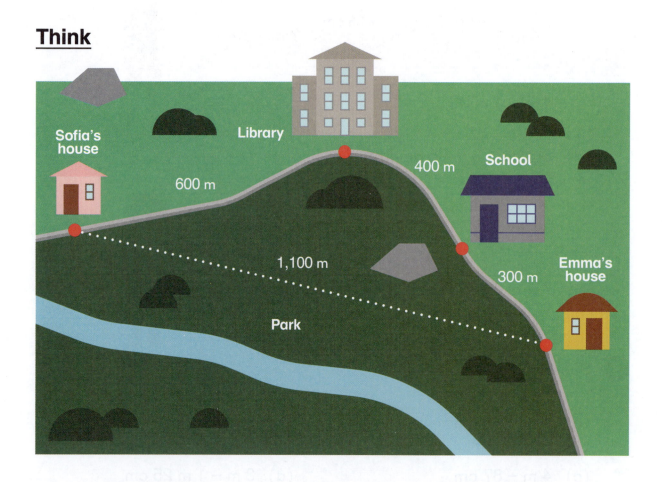

(a) Sofia biked from her house to the library and then to the school.
How far did she bike?

(b) Then, Sofia biked from the school to Emma's house.
How far did she bike in all?

(c) Sofia biked home from Emma's house by going straight through the park.
How far did she bike on the way home?

Which route was shorter?
How much shorter?

Learn

(a)

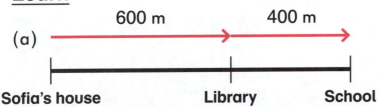

600 m + 400 m = 1,000 m

Sofia biked [____] m to get from her house to the library to the school.

> The **kilometer** is a unit of length.
> We write **km** for kilometer.
> 1 km = 1,000 m

We can express this distance using kilometers.

(b)

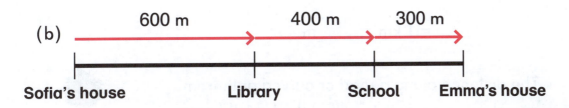

600 m + 400 m + 300 m = 1,300 m
$\qquad\qquad\qquad\qquad$ = 1 km 300 m

1,300 m
1,000 m   300 m

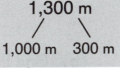

Sofia biked [____] km [____] m to Emma's house.

(c) 1,100 m = [____] km [____] m

Sofia biked [____] km [____] m
to go home through the park.

It is 200 m shorter to go through the park.

Do

1

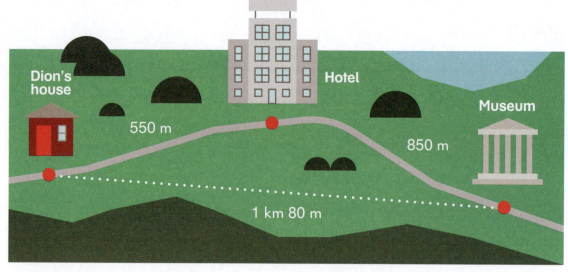

(a) Find the travel distance from Dion's house to the hotel and then to the museum in kilometers and meters.

550 m + 850 m = ⬚ m

= 1 km ⬚ m

> The distance along a road or curved path from one place to another is called the **travel distance**.

(b) Find the direct distance from Dion's house to the museum in meters.

1 km + 80 m = 1,000 m + ⬚ m

= ⬚ m

> The distance along a straight line from one place to another is called the **direct distance**.

2 The distance from Emma's house to her grandmother's house is 3,500 m.
Express this distance in kilometers and meters.

3,500 m = 3,000 m + [ ] m

= [ ] km [ ] m

1,000 m = 1 km
3,000 m = ? km

3 The distance from the school to the amusement park is 4 km 50 m.
Express this distance in meters.

4 km 50 m = 4,000 m + [ ] m

= [ ] m

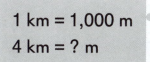

1 km = 1,000 m
4 km = ? m

4 Write in kilometers and meters.

(a) 6,700 m

(b) 6,070 m

(c) 6,007 m

(d) 1,275 m

(e) 8,997 m

(f) 4,022 m

5 Write in meters.

(a) 5 km 300 m

(b) 5 km 30 m

(c) 5 km 3 m

(d) 1 km 375 m

(e) 8 km 8 m

(f) 6 km 28 m

Exercise 3 • page 88

Think

Kona drove from her house to the post office and then to the hospital.

She drove a total distance of 2 km.

How far is the hospital from the post office?

Learn

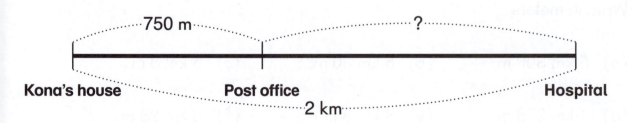

2 km − 750 m = ?

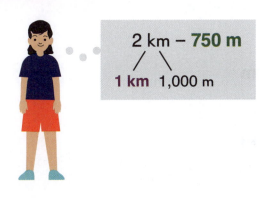

2 km − **750 m**
        /   \
1 km 1,000 m

1,000 m — **−750 m** → 250 m — **+1 km** → [____] km [____] m

Method 2

1 km = 750 m + 250 m
1 km is **250 m** more than 750 m.

2 km — **−1 km** → 1 km — **+250 m** → [____] km [____] m

Method 3

2,000 m − 750 m = 1,250 m

1,250 m = [____] km [____] m

2 km = 2,000 m
2,000 − 750 =?

The hospital is [____] km [____] m from the post office.

Do

1 Find pairs of lengths that make 1 km.

380 m	405 m	670 m	462 m
595 m	330 m	620 m	701 m
299 m	60 m	538 m	940 m

1,000 = 9 hundreds + 9 tens + 10 ones

2 (a) Subtract 610 m from 1 km.

1 km − 610 m = ⬛ m

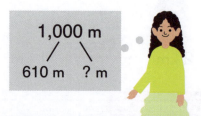

1,000 m
/ \
610 m   ? m

(b) Subtract 610 m from 3 km.

3 km − 610 m = ⬛ km ⬛ m

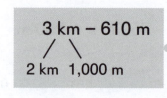

3 km − 610 m
/ \
2 km   1,000 m

3 Subtract 25 m from 2 km.

2 km − 25 m = ⬛ km ⬛ m

4 Subtract.

(a) 4 km − 1 km 300 m

4 km —**− 1 km**→ 3 km —**− 300 m**→ ☐ km ☐ m

(b) 7 km − 3 km 850 m

7 km —**− 3 km**→ ☐ km —**− 850 m**→ ☐ km ☐ m

5 Subtract in compound units.

(a) 1 km − 260 m   (b) 1 km − 60 m

(c) 4 km − 830 m   (d) 3 km − 1 km 330 m

(e) 8 km − 4 km 420 m   (f) 20 km − 10 km 500 m

6 (a) 1 km = 850 m + ☐ m

(b) 4 km 150 m + ☐ m = 5 km

(c) 3 km = 2 km 5 m + ☐ m

(d) 6 km = 4 km 750 m + 1 km ☐ m

Exercise 4 • page 91

Think

How can we use beakers to measure the capacity of a paper cup or a bottle in liters and milliliters?

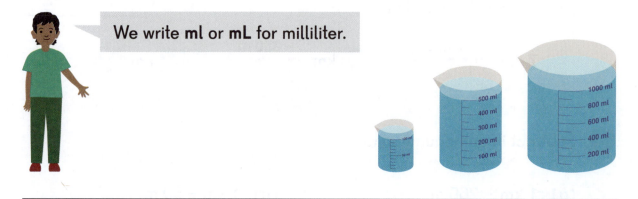

We write **ml** or **mL** for milliliter.

Fill a small paper cup with water.
Estimate the capacity of the cup in milliliters.
Then measure its capacity using a 500-mL beaker.

Fill a water bottle with water.
Estimate the capacity of the bottle in milliliters.
Then measure its capacity using a 1,000-mL beaker.

Learn

(a)

The tick marks show increments of 50 mL. This cup holds a little more than 100 mL.

(b)

The tick marks show increments of 100 mL. This water bottle holds about 500 mL.

The **capacity** of a container is the amount of liquid it can hold.

> The amount of liquid in a container is the **volume** of the liquid.
> We measure liquid volume in **liters** (L) and **milliliters** (mL).
> 1 L = 1,000 mL

Which beaker has a capacity of 1 L?

A teaspoon holds about 5 mL of water.

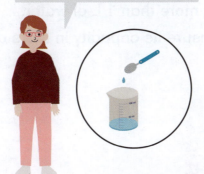

A 1-cm cube would hold 1 mL of water.

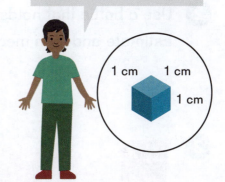

Do

1 1 water bottle has a capacity of 500 mL.

What is the total capacity of 2 water bottles in liters?

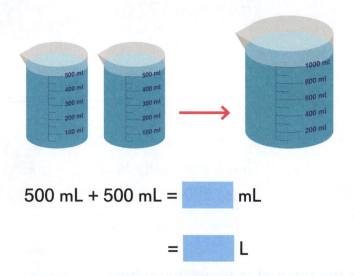

500 mL + 500 mL = ⬜ mL

= ⬜ L

Milli- is used for 1 out of 1 thousand.
There are 1 thousand milliliters in 1 liter.

2 Use a bottle that holds more than 1 L of water.

Estimate and then measure its capacity in liters and milliliters.

3

The capacity of the carton is ⬜ mL.

4

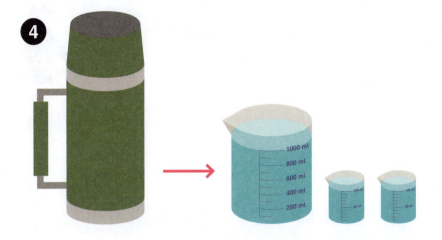

(a) The capacity of the thermos is [   ] mL.

(b) Its capacity in liters and milliliters is [   ] L [   ] mL.

5

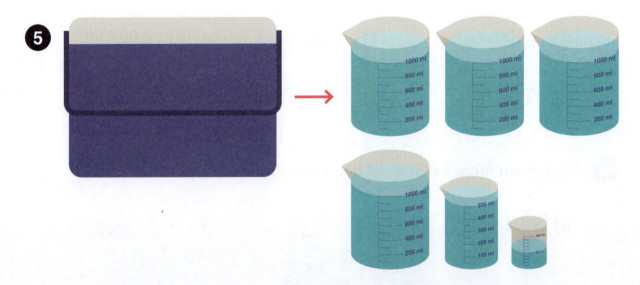

(a) What is the capacity of the cooler in milliliters only?

(b) What is the capacity of the cooler in liters and milliliters?

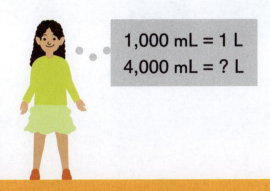

1,000 mL = 1 L
4,000 mL = ? L

6 Write 2 L 50 mL in milliliters.　　2,000 mL + 50 mL = ? mL

2 L 50 mL = ▭ mL

7 Write in milliliters.

(a) 3 L 500 mL　　　(b) 3 L 50 mL　　　(c) 3 L 5 mL

(d) 4 L 25 mL　　　(e) 6 L 8 mL　　　(f) 4 L 250 mL

8 Write in liters and milliliters.

(a) 2,400 mL　　　(b) 2,040 mL　　　(c) 2,004 mL

(d) 6,025 mL　　　(e) 6,205 mL　　　(f) 1,750 mL

9 How much more water must be added to make...

(a) 1 L?

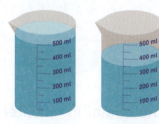

(b) 2 L?

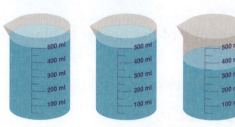

Exercise 5 • page 94

Think

Emma and Sofia are using a platform scale to weigh some objects in kilograms and grams.

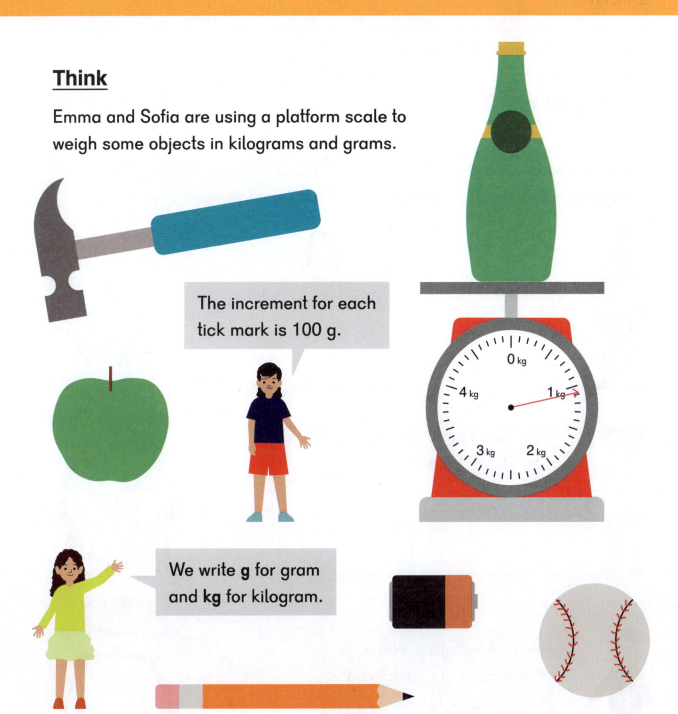

The increment for each tick mark is 100 g.

We write **g** for gram and **kg** for kilogram.

Find some objects that you think weigh between 1 kg and 5 kg.
About how many kilograms do you think each object weighs?
Measure their weights on a platform scale.
Write the weights in kilograms and grams.

Learn

The heavier the object, the more mass it has.
We measure mass in **kilograms** (kg) and **grams** (g).
1,000 g = 1 kg

Kilo- is used for thousands.
1 kilogram is 1 thousand grams.

1 mL of water weighs 1 g.
1 L of water weighs 1 kg.

This potted plant weighs 1 kg 400 g.
1 kg 400 g = 1,400 g

This rock weighs 1,600 g.
Where would the needle on
this scale point?

Write the weights of the objects you weighed in grams.

Do

1 Put different combinations of weights from a weight set on the scale to make 1 kg.

2 Read the scales.
Write the weight in both kilograms and grams.

3 A bag of potatoes weighs 3 kg 45 g.
Write the weight in grams.

4 A bag of flour weighs 1 kg 340 g.
A bag of sugar weighs 2,230 g.
Which is heavier, the bag of flour or the bag of sugar?

5 Write in grams.

(a) 3 kg 630 g   (b) 7 kg 455 g   (c) 5 kg 593 g

(d) 6 kg 55 g   (e) 4 kg 2 g   (f) 8 kg 8 g

6 Write in kilograms and grams.

(a) 2,400 g   (b) 6,115 g   (c) 2,764 g

(d) 3,305 g   (e) 4,027 g   (f) 7,003 g

7 How many more grams do we need on the left side of the scale to make...

(a) 1 kg?

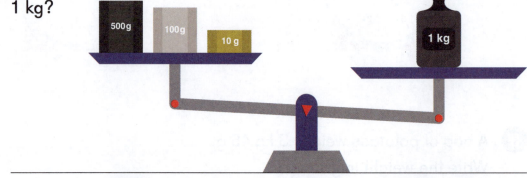

(b) 2 kg?

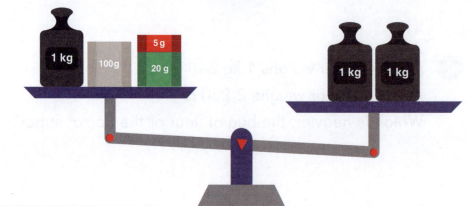

Think

The distance from the bakery to the hotel is 2 km 750 m.

The hotel is 800 m closer to Alex's house than it is to the bakery.

How far is the bakery from Alex's house?

Give your answer in compound units.

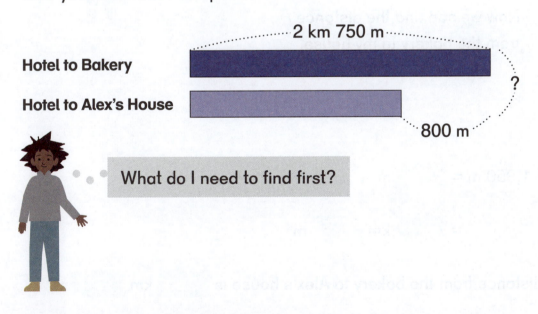

Hotel to Bakery

Hotel to Alex's House

What do I need to find first?

Learn

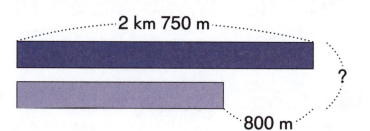

Hotel to Bakery

2 km 750 m

Hotel to Alex's House

?

800 m

I need to first find the distance from the hotel to Alex's house.

2 km 750 m = 2,750 m

2,750 m − 800 m = 1,950 m

= ⬜ km ⬜ m

$$\begin{array}{r} 2{,}750 \\ -\ \ 800 \end{array}$$

Now we can find the distance from the bakery to my house.

2,750 m + 1,950 m = ⬜ m

= ⬜ km ⬜ m

The total distance from the bakery to Alex's house is ⬜ km ⬜ m.

Do

1 Dion has a board that is 2 m 20 cm long.
He cuts off 90 cm to make a ramp for his remote control car.

(a) What is the length of the remaining piece of board?

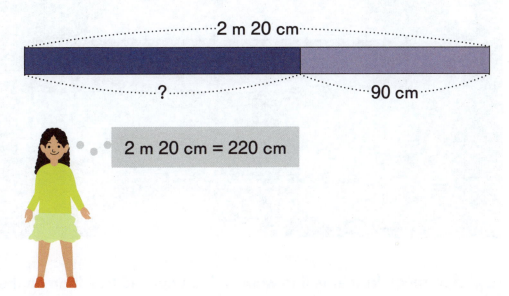

2 m 20 cm = 220 cm

(b) What is the difference in length between the two pieces of board?

2 1 bottle of water weighs 520 g.
What do 3 bottles of water weigh in kilograms and grams?

3 × 520 = ?

3 There are two trails to the scenic overlook.
The paved trail is 950 m.
The unpaved trail is 2 km 200 m.

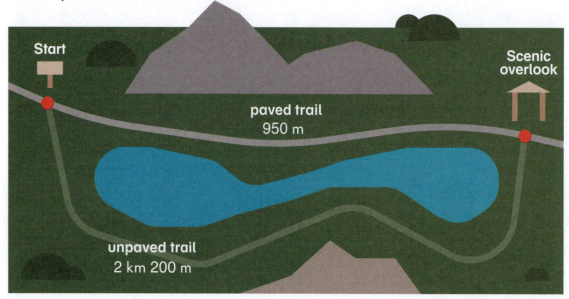

(a) How much farther is it to walk on the unpaved trail than on the paved trail in kilometers and meters?

(b) Sofia went to the scenic overlook on the unpaved trail, and came back on the paved trail.
How far did she walk?

4 How much do the can of paint and the bucket of plaster weigh altogether?

5 kg 930 g is 5,930 g.

$$\begin{array}{r} 5,930 \\ +\ 3,685 \\ \hline \end{array}$$

5 kg 930 g   3,685 g

5 A jar full of jelly beans weighs 1 kg 500 g.

The empty jar weighs 350 g.

What do the jelly beans weigh?

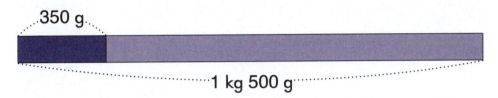

6 Mr. Kiper has 1 L 250 mL of solution in a flask.

After pouring the same amount into each of 5 beakers,

he has 550 mL left.

How much solution did he pour into each beaker?

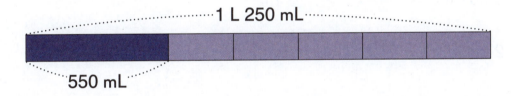

7 A bag of nails is 4 times as heavy as a hammer.

The hammer weighs 945 g.

How much do the bag of nails and the hammer weigh altogether?

Give your answer in kilograms and grams.

Nails

Hammer

945 g

1 (a) 159 cm = ▢ m ▢ cm

(b) 408 cm = ▢ m ▢ cm

(c) 1 m 98 cm = ▢ cm

(d) 2,034 m = ▢ km ▢ m

(e) 8 km 9 m = ▢ m

(f) 5,001 mL = ▢ L ▢ mL

(g) 2 L 432 mL = ▢ mL

(h) 3,215 g = ▢ kg ▢ g

(i) 8 kg 82 g = ▢ g

2 Put the lengths in order from shortest to longest.

| 4 m 5 cm | 54 cm | 4 m 50 cm | 540 cm |

3 Put the weights in order from lightest to heaviest.

| 6 kg 5 g | 6,050 g | 5,600 g | 6 kg 500 g |

4 (a) 1 m − 73 cm = ▭ cm

(b) 2 m − 9 cm = ▭ m ▭ cm

(c) 3 m − 1 m 5 cm = ▭ m ▭ cm

(d) 1 km − 620 m = ▭ m

(e) 4 km − 90 m = ▭ km ▭ m

(f) 5 km − 2 km 850 m = ▭ km ▭ m

(g) 1 L − 405 mL = ▭ mL

(h) 4 L − 35 mL = ▭ L ▭ mL

(i) 6 kg − 120 g = ▭ kg ▭ g

(j) 7 kg − 5 g = ▭ kg ▭ g

5

The gift shop is how much closer to the bakery than to the factory?

6 The capacity of a canteen is 1 L 500 mL.
The capacity of a thermos is 655 mL.

(a) What is the total capacity of the canteen and thermos in liters and milliliters?

(b) How much more water can the canteen hold than the thermos?

7 Sebastian is 1 m 35 cm tall.
His brother is 96 cm tall.
Who is taller and by how much?

8 A plank of wood is 5 m 85 cm long.
It is cut into 9 pieces of equal length.
How long is each piece?

9 Hudson walked a total of 950 m along the river trail and then back again.
How far did he walk in kilometers and meters?

10 Susma ran around a 515-m track 4 times.
How far did she run in kilometers and meters?

11 A baker has 3 kg of flour.
She uses 900 g to make muffins and 700 g to make pastries.
How many kilograms and grams of flour does she have left?

Exercise 8 · page 105

1 Estimate, then find the value.

(a) 2,306 + 3,895

(b) 7,329 + 494

(c) 6,475 + 1,785

(d) 4,926 − 1,469

(e) 8,152 − 267

(f) 5,003 − 1,308

2 Solve.

(a) 87 × 3

(b) 45 × 7

(c) 95 ÷ 6

(d) 57 ÷ 4

(e) 329 ÷ 5

(f) 268 × 9

(g) 843 × 8

(h) 915 ÷ 8

3 Which of the following shapes have $\frac{3}{5}$ colored?

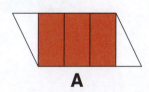

A

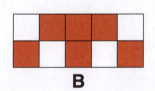

B

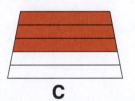

C

D

4

```
├──┼──┼──┼──┼──┼──┼──┼──→
0    A    B    C    1    D    E
```

(a) What number is indicated by C?

(b) Which letter indicates the number $\frac{5}{4}$?

(c) How many eighths make the number indicated by B?

5 What sign, >, <, or =, goes in the ◯?

(a) $\frac{5}{6}$ ◯ $\frac{5}{8}$

(b) $\frac{4}{8}$ ◯ $\frac{7}{16}$

(c) $\frac{3}{5}$ ◯ $\frac{1}{3}$

(d) $\frac{2}{3}$ ◯ $\frac{4}{7}$

(e) $\frac{3}{7} + \frac{2}{7}$ ◯ $\frac{7}{8} - \frac{2}{8}$

(f) $\frac{11}{12} - \frac{5}{12}$ ◯ $\frac{3}{8} + \frac{3}{8}$

6 (a) The volume of water in this beaker
is ⬜ L ⬜ mL.

(b) 750 mL of water is poured out.
How much water is still in the beaker?
Give your answer in compound units.

5 L

7 This bar graph shows the length of five different hikes in kilometers.

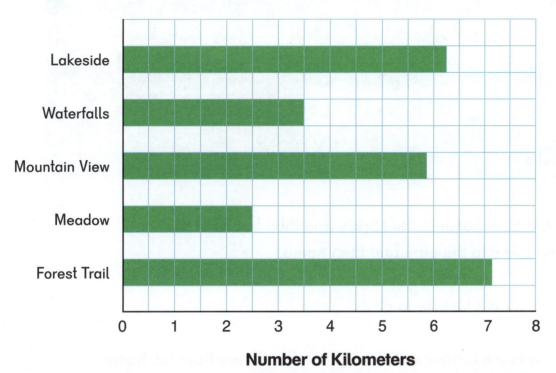

Number of Kilometers

(a) Each square on the graph shows increments of ⬜ m.

(b) List the hikes in order from shortest to longest.

(c) Which two hikes are about the same length?

(d) Dion went on the hike that was 6 km 200 m.
Which hike did he go on?

(e) Sofia went on two hikes.
She hiked about 8 km.
Which two hikes did she go on?

8

Jett biked from home to the library and then back home.
Next, he biked to the park and then home.

(a) How far did he bike in all?

(b) How much farther is the park than the library from his home?

9 Sara found 4 times as many seashells as Dana.
Camila found twice as many seashells as Dana.
Altogether, the 3 friends found 98 seashells.
How many more seashells did Sara find than Camila?

10 The sum of two numbers is 770.
The difference between them is 200.
What is the greater number?

11 8 identical boxes weigh 768 g.
2 identical crates weigh 980 g.
How much do 1 crate and 1 box weigh altogether?

Exercise 9 • page 109

Chapter 12

Geometry

Make pictures with tangrams.

Lesson 1
Circles

Think

Fold a paper circle in half and unfold.
Fold it in half a different way and unfold again.

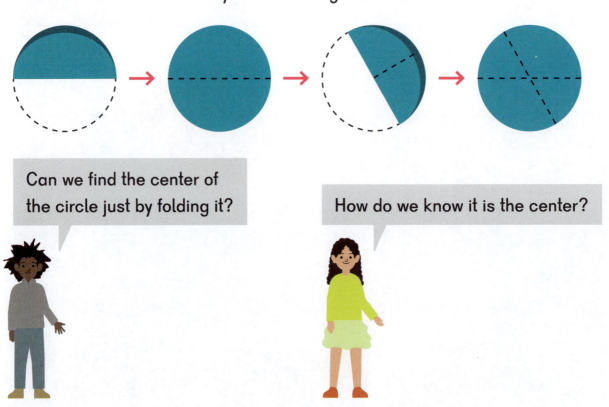

Can we find the center of the circle just by folding it?

How do we know it is the center?

Use a ruler to measure the distance from the center of the circle to different points on the edge of the circle.

(a) What can you say about the distance from the center of the circle to any point on the edge of the circle?

(b) How long is the distance from the center of the circle to the edge of the circle compared to the distance across the circle through the center?

Learn

The edge of a **circle** is always the same distance from its center.

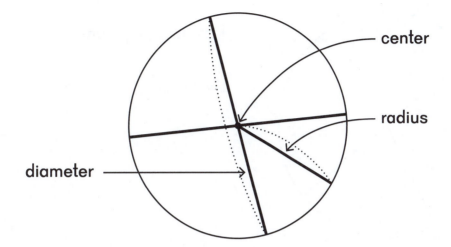

A **radius** of a circle is a straight line from the center of the circle to its edge.

All **radii** of a circle have the same length.

The **diameter** of a circle is a straight line that goes all the way across the circle through the center.

All diameters of a circle have the same length.

All diameters intersect at the center of the circle.

The diameter of a circle is [ ] times as long as the radius of the circle.

Do

1 Measure the radius and diameter of the circle below in centimeters.

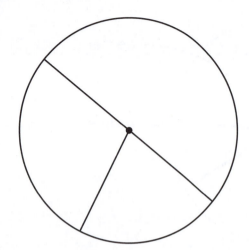

Radius: [____] cm

Diameter: [____] cm

2 (a) How long is the diameter of each circle?

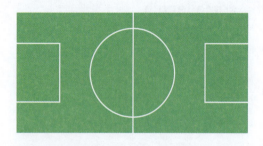 

Circle in the center of soccer field
Radius: 30 ft

Mirror
Radius: 4 in

(b) How long is the radius of each circle?

Trampoline
Diameter: 120 in

Bicycle Wheel
Diameter: 12 in

3 How long is the radius of each circle?

(a)

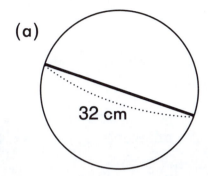

32 cm

(b)

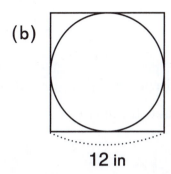

12 in

4 These quarter-circles were each cut from a whole circle.
How long is the diameter of each original circle?

(a)

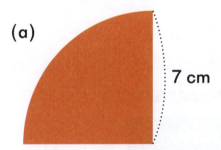

7 cm

(b)

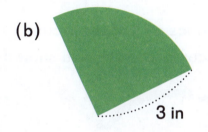

3 in

5 The two smaller circles have the same radii.
How long is the diameter of the largest circle?

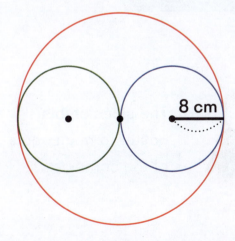

8 cm

Exercise 1 • page 115

Think

Fasten two cardboard strips at one end with a brad.
Open it to make an angle.

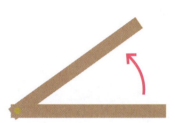

You can make different angles by opening the sides by different amounts.

What shapes do you see around you that have angles?
Use the cardboard strips to show these angles.

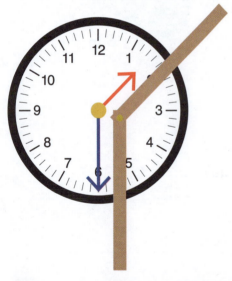

The hands of this clock form an angle.

The sides of this poster form an angle at the corner.

Learn

Two lines (sides) that meet at a point form an **angle**.

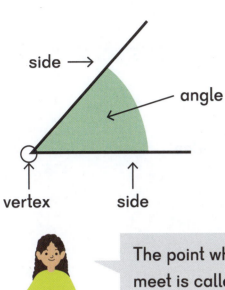

side →

angle

vertex

side

The angle is the space between the sides.

The point where the lines meet is called the **vertex**.

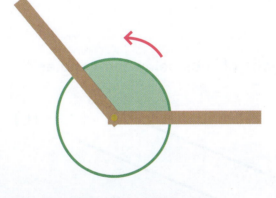

An angle is formed when one side opens away from the other side around a circle. The larger the opening, the larger the angle.

Do

1 (a)

A rectangle has [ ] angles and [ ] sides.

(b)

A triangle has [ ] angles and [ ] sides.

2 Which pairs of lines below form angles?

(a)

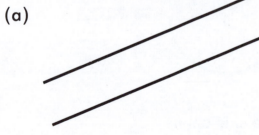

(b)

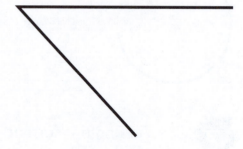

(c)

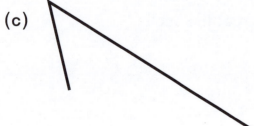

(d)

3 Use the corners of set squares to draw angles.

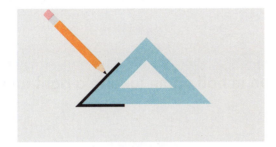 

4 Use the cardboard strips to draw different angles.

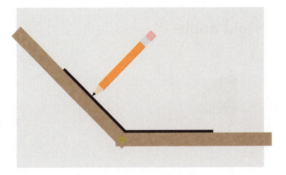 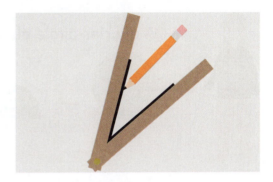

5 Use a geoboard to make a shape that has...

(a) 5 angles.

(b) 6 angles.

(c) 8 angles.

How many sides do each
of your shapes have?

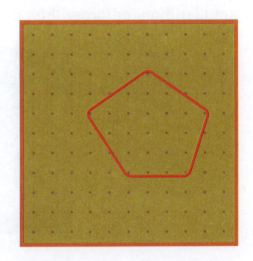

Exercise 2 • page 119

Think

Fold a torn piece of paper in half and then in half again to form a right angle.

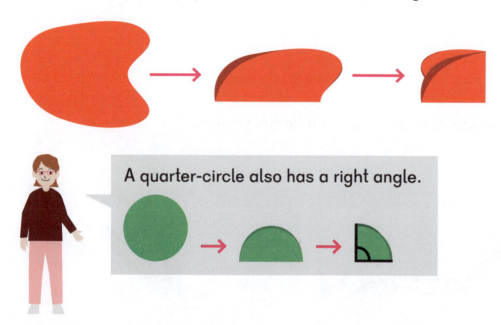

A quarter-circle also has a right angle.

What shapes do you see around you that have right angles?
Use the right angle to find out if the angles around you are the same as, smaller than, or larger than a right angle.

MATH

COOL

Use the cardboard strips to form angles that are the same as, smaller than, or larger than a right angle.

Learn

These angles are **right** angles.

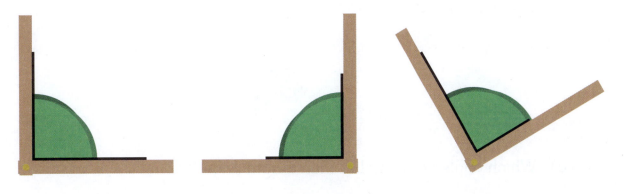

These angles are smaller than right angles.

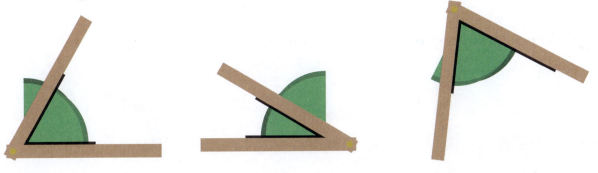

These angles are larger than right angles.

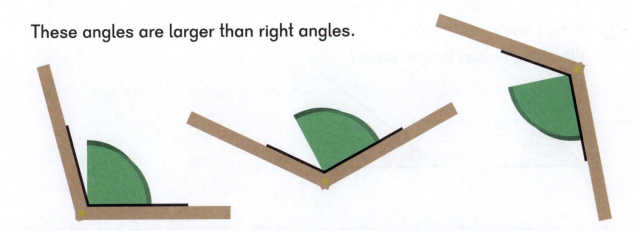

The size of the angle depends on the size of the opening between the sides, not the length of the sides.

Do

1 (a) Which angles on the set squares are right angles?
Use your folded paper to find out.

(b) Which angles are smaller than a right angle?

(c) Which angles are larger than a right angle?

2 Compare the angles of the set squares.

3 Which angle is larger?
Which angle has longer sides?

 a 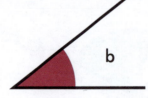 b

4 Use a set square or your folded circle to examine the angles on tangram pieces.
Which pieces have right angles?

5 (a) Which one of these angles is a right angle?

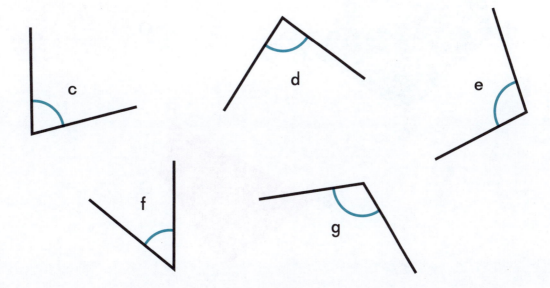

(b) List the angles in order from smallest to largest.

6

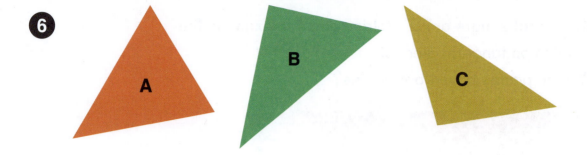

(a) Which triangle has a right angle?

(b) Which triangle has an angle larger than a right angle?

(c) Which triangle has all angles smaller than a right angle?

7 Find the right angles in the shapes below.

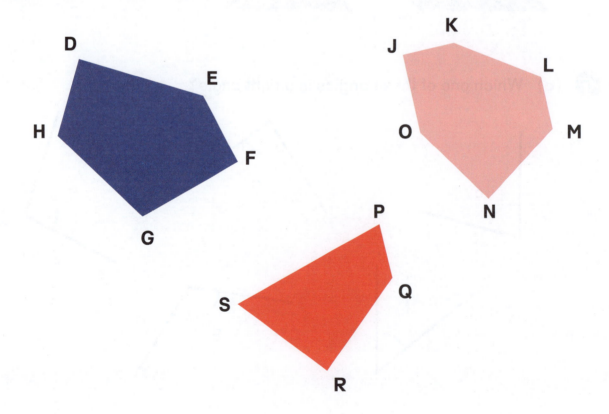

Exercise 3 · page 121

12-3 Right Angles

Think

Use sticks to make different triangles.
How can we group the triangles according to the length of the sides?

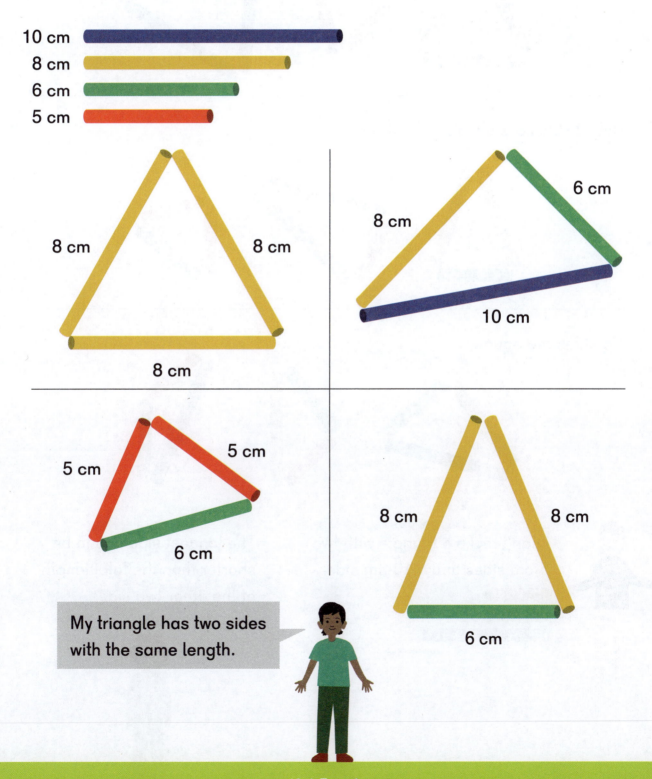

10 cm

8 cm

6 cm

5 cm

8 cm 8 cm

8 cm

8 cm 6 cm

10 cm

5 cm 5 cm

6 cm

8 cm 8 cm

6 cm

My triangle has two sides with the same length.

Learn

There are three kinds of triangles based on side lengths.

All 3 sides are equal.

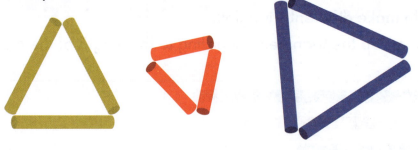

Only 2 sides are equal.

No sides are equal.

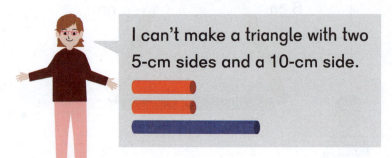

I can't make a triangle with two 5-cm sides and a 10-cm side.

The longest side has to be shorter than the total length of the other two sides.

Do

1 Measure the sides of these triangles in centimeters.

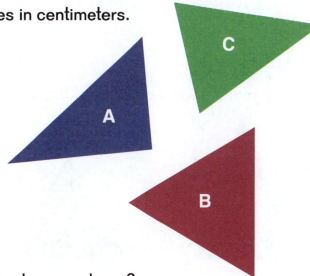

(a) Which sides on each triangle are equal?

(b) Which triangle has a right angle?

2 How many equal sides does each set square have?

3 Equal sides are marked on these triangles. How many equal sides does each triangle have?

(a)

(b)

(c)

(d)

We draw the same number of marks on sides that are the same length.

Exercise 4 · page 125

Think

How many angles on each of these triangles are equal?

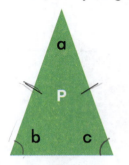

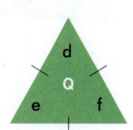

Learn

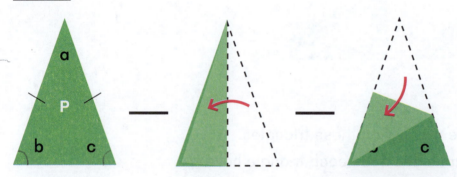

☐ angles are the same size.

The angles opposite equal sides are equal.

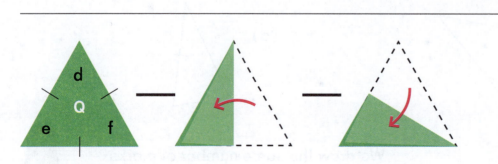

☐ angles are the same size.

Do

1 Put two set squares together to make triangles.
Trace around them.

(a) How many different triangles can you make?

(b) Which sides of each triangle have the same length?

(c) Which angles of each triangle are the same size?

(d) Which triangle is a right triangle with two equal sides?

2 Use circle dot paper to draw triangles by connecting the dots on the edge of the circles.

(a) Draw some triangles with three equal sides.
Which angles are equal?

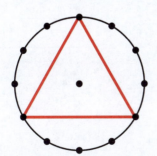

(b) Draw some triangles with two equal sides.
Which angles are equal?

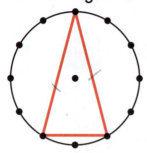

Do any of them
have a right angle?

(c) Draw some triangles with no equal sides.

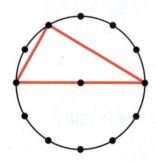

Do any of them
have a right angle?

3 Draw triangles using the center of the circle for one vertex.
What is the same about all of them?

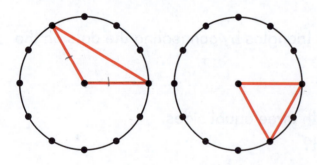

Two of the sides are radii...

Exercise 5 • page 127

Think

Make different **quadrilaterals** on a geoboard.

A quadrilateral has 4 straight sides.

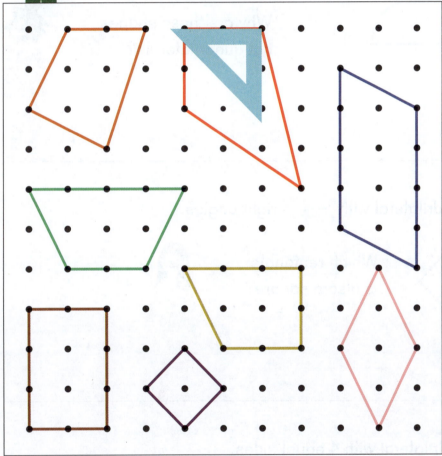

Use the right angle on a set square to check the size of the angles.

What do you notice about the sides and angles of each of your quadrilaterals?

Do any of them have right angles?

Do any of them have equal sides?

Do any of them have equal angles?

Learn

This quadrilateral has ▮ right angle,
▮ angles greater than a right angle,
and ▮ angle smaller than a right angle.

It has ▮ equal sides.

These shapes are not quadrilaterals.

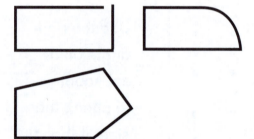

Why are these shapes not quadrilaterals?

A rectangle is a quadrilateral with ▮ right angles.

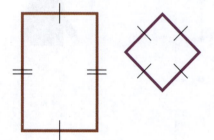

Which rectangle is also a square?

A **rhombus** is a quadrilateral with 4 equal sides.

 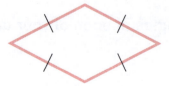

A square is also a rhombus.

Do

1 Put tangram pieces together to make quadrilaterals.

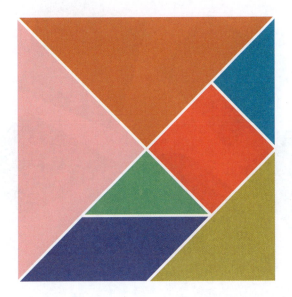

2 Put two set squares together to make quadrilaterals.
Trace around them and cut them out.

What can you say about the opposite angles in a rhombus?

(a) Which quadrilaterals have equal sides?

(b) Which quadrilaterals have equal angles?

(c) Are any of them rhombuses?

3 Which of the following figures are rhombuses?
Measure the sides in centimeters.

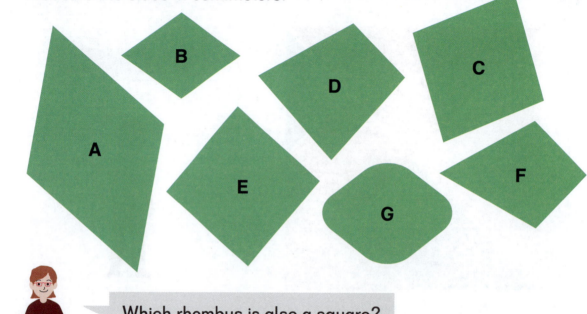

Which rhombus is also a square?

4 Use dot paper or graph paper to draw each quadrilateral below.

(a) A quadrilateral with 4 right angles.

(b) A quadrilateral with 2 angles smaller than a right angle.

(c) A quadrilateral with at least 1 of each kind of angle.

Can all 4 angles on a quadrilateral be greater than a right angle?

Exercise 6 • page 130

Think

A **compass** is a tool that is used to draw circles.
There are different kinds of compasses.

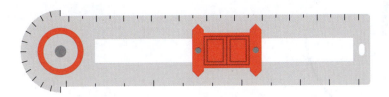

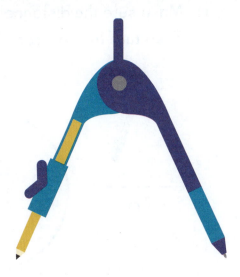

How can we use a compass to draw circles of a given radius?

Can we use a compass to help us draw triangles?

(a) Draw a circle with a radius of 5 cm.

(b) Draw a circle with a diameter of 8 cm.

What is the radius?

(c) Use a compass and a ruler to draw a triangle with sides that measure 3 cm, 4 cm, and 4 cm.

Draw the 3 cm side first.

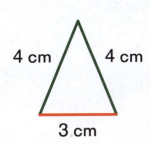

4 cm   4 cm

3 cm

How can I use the compass to draw the other two sides?

Learn

(a) Make sure the distance from the center to the pencil point is 5 cm.
Then turn the compass to draw the circle.

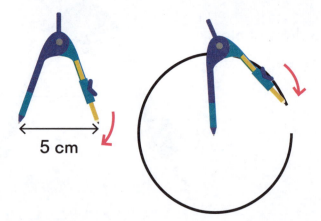

(b) The distance from the center to the pencil point should be 4 cm to draw a circle with a diameter of 8 cm.

(c) Draw the 3 cm side.
Then use the compass to find out where to put the third vertex.

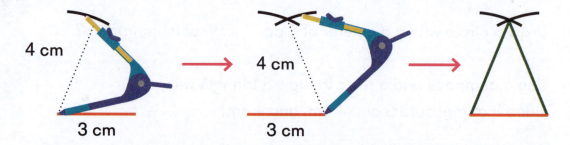

How do we know that the length of each of the two other sides is 4 cm?

Do

1 Draw the following circles with a compass.

> Use a ruler or centimeter graph paper.

(a) A circle with a radius of 6 cm.

(b) A circle with a diameter of 10 cm.

(c) A circle with a radius of 4 cm, then a second circle with the same center and a radius of 6 cm.

2 Draw a triangle with sides that are all 4 cm long.

3 (a) Dion drew a triangle with sides of lengths 5 cm, 4 cm, and 3 cm using centimeter graph paper, a straight edge, and a compass. He started by drawing the 5-cm side.

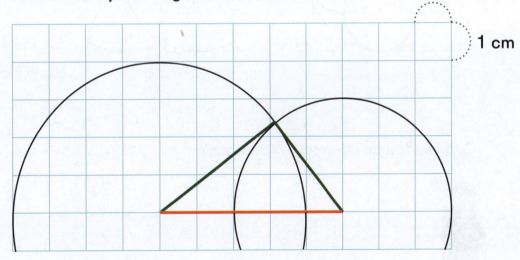

1 cm

1 cm

Which of the angles is a right angle?

(b) Draw a triangle with sides that are 4 cm, 5 cm, and 3 cm long. Start with the 4 cm side.

4 cm

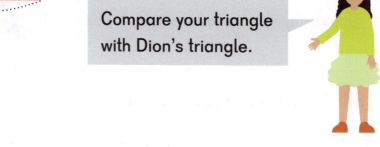

Compare your triangle with Dion's triangle.

4 Draw a triangle with sides that are 6 cm, 7 cm, and 8 cm long.

5 Mei drew these designs on centimeter graph paper using a compass. Draw some other circle designs using these tools.

1 cm

1 cm

Can you tell how I drew my designs?

Exercise 7 · page 134

12-7 Using a Compass

1 How long is the radius and the diameter of each circle?

(a)

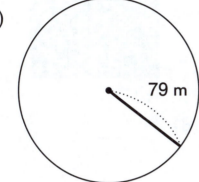

79 m

(b)

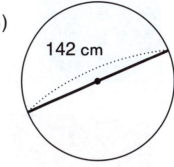

142 cm

2 List the angles in order from smallest to largest.

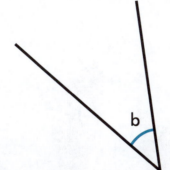

a

b

c

d

3 How many angles are the same size in a triangle with...

(a) only 2 equal sides?

(b) 3 equal sides?

(c) no equal sides?

4 How many of each type of angle (equal to, greater than, or less than a right angle) do each of these figures have?

(a)

(b)

(c)

(d)

5

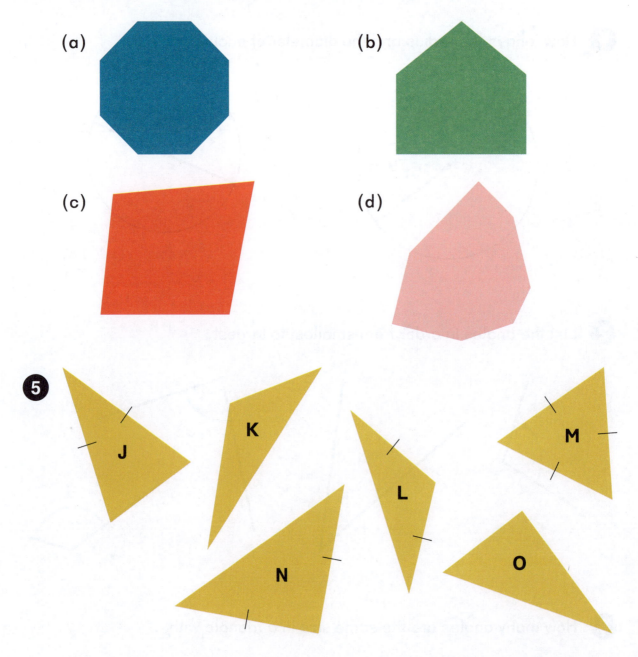

(a) Which triangles have a right angle?

(b) Which triangles have only 2 angles the same size?

(c) Which triangle has 3 angles the same size?

6

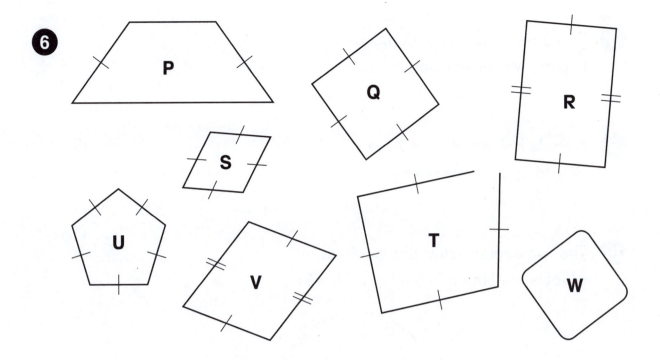

(a) Which of these figures are quadrilaterals?

(b) Which of them are rectangles?

(c) Which of them are rhombuses?

7 (a) Use circle dot paper to draw different quadrilaterals by connecting the dots on the edge of the circles or the center.

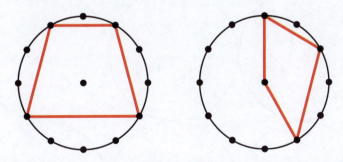

(b) Draw a rectangle.

(c) Draw a rhombus.

8 Can a rectangle be a rhombus?
Explain why or why not.

9 A circle has a radius of 7 cm.
How long is its diameter?

10 The two circles below are identical.
What is the length of the longer line?

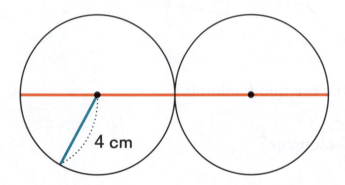

4 cm

11 Each of these circles has a radius of 3 cm.
What is the length and width of the rectangle?

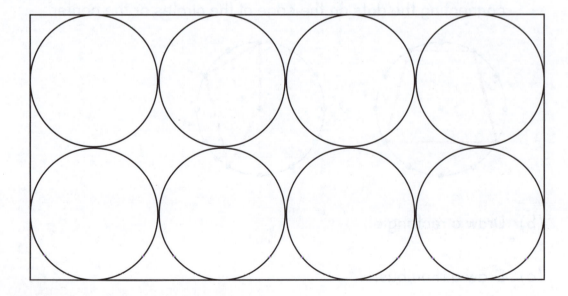

Exercise 8 • page 139

12-8 Practice

Chapter 13

Area and Perimeter

Think

Which shape covers more surface, the square or the rectangle?

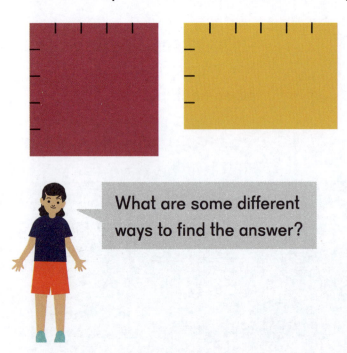

What are some different ways to find the answer?

Learn

Lay one paper on top of the other and cut off the extra piece.

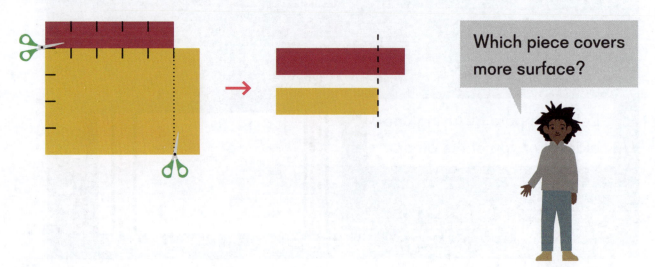

Which piece covers more surface?

Lay equal-sized squares on top of each shape.

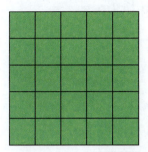 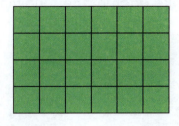

How many squares does it take to cover each shape?

> The size of a closed surface is called the **area**.
> Area is measured in **square units**.

The area of the square is ▢ square units.

The area of the rectangle is ▢ square units.

The _____ has a greater area than the _____.

Do

1 Use square cards or tiles to make each of the following figures.

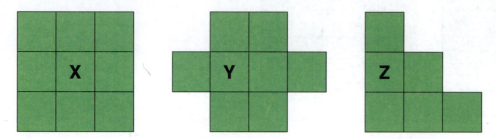

(a) How many square units is the area of each figure?

(b) Which figure has the largest area?

(c) Which figure has the smallest area?

2 Make different figures with 5 square cards or tiles.
What can you say about the area of each figure?

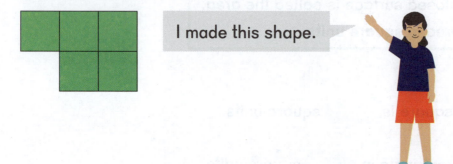

I made this shape.

3 Use 4 square cards and 4 half-square cards to make different figures.
What is the area of each figure?

I made this shape.

4 (a) Find the area of each figure in square units.

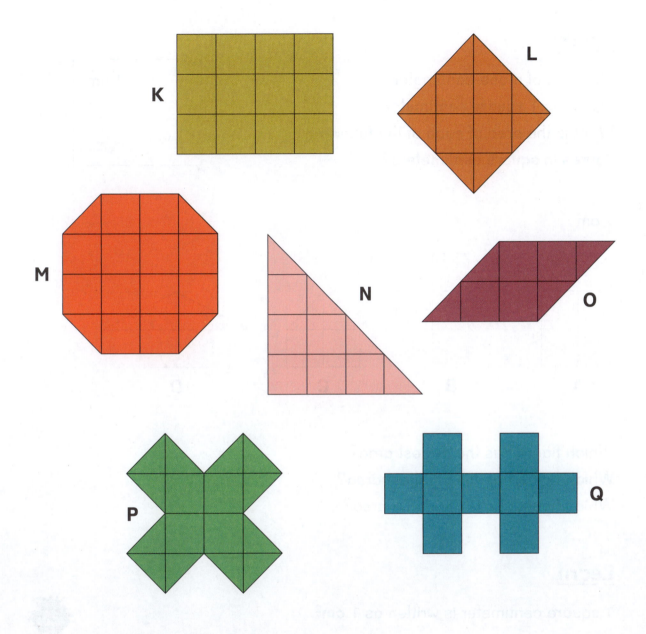

(b) Which figure has the largest area?

(c) Which figure has the smallest area?

(d) Which figures have the same area?

Exercise 1 • page 143

Think

The area of a one-centimeter square is 1 **square centimeter**. What is the area of each of the following figures in square centimeters?

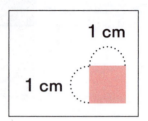

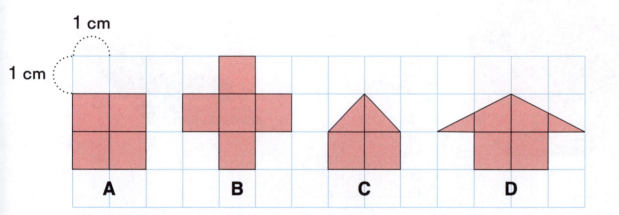

Which figure has the largest area?
Which figure has the smallest area?
Which figures have the same area?

Learn

1 square centimeter is written as 1 **cm²**.

Figure	A	B	C	D
Area	4 cm²	5 cm²	3 cm²	4 cm²

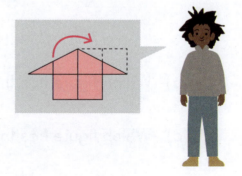

Figure _____ has the largest area.

Figure _____ has the smallest area.

Figures _____ and _____ have the same area.

Do

1 Use one-centimeter graph paper to draw...

(a) three figures that have an area of 6 cm².

(b) three figures that have an area of 8 cm².

2 What is the area of each of the following figures?

1 cm

1 cm

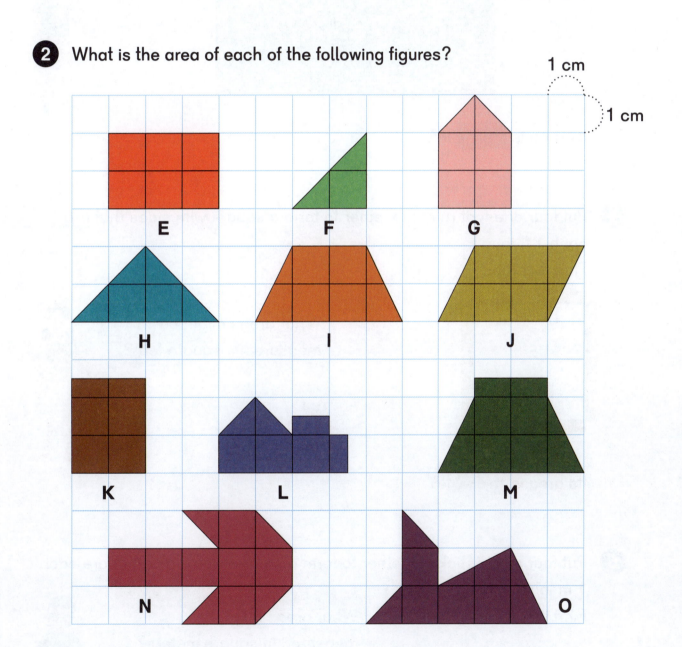

3 The figure below is made of one-inch squares.
Find its area.

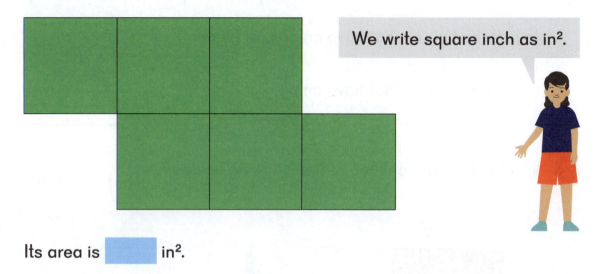

We write square inch as in².

Its area is [ ] in².

4 Put four one-foot rulers together to form a square with sides that are each 1 ft long.
Find its area.

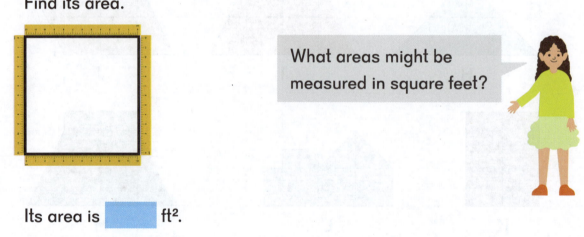

What areas might be measured in square feet?

Its area is [ ] ft².

5 Put four meter sticks together to form a square with sides that are each 1 m long.
Find its area.

What areas might be measured in square meters?

Its area is [ ] m².

Exercise 2 • page 146

Lesson 3
Area of Rectangles

(3)

Think

A rectangular photo is 6 in long and 4 in wide.
Find its area.

6 in

4 in

How many one-inch squares do we need to cover the whole rectangle?

Can we use the measurements for the side lengths?

Learn

How many groups of 4 squares do I need to cover the rectangle?

6 × 4 =

Area: in²

How many groups of 6 squares do I need to cover the rectangle?

4 × 6 =

6 × 4 = 4 × 6

Area =   in²

Area of rectangle = length × width or width × length

<u>Do</u>

1 Make different rectangles using 12 square tiles.

    (a) What is the length and width of each rectangle you made?

    (b) What is the area of each rectangle you made?

2 Make a square using more than 4 square tiles.
What is the area of the square?

3 Find the area of each rectangle.

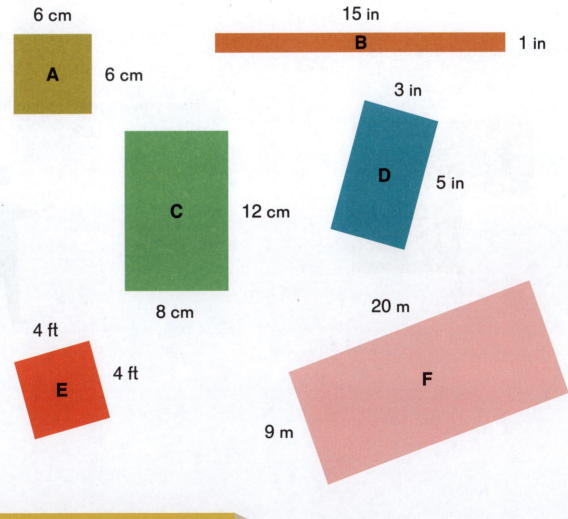

Think

We are putting these shapes together to make colorful pictures. How can we find the area of each shape?

Find the area of this figure.

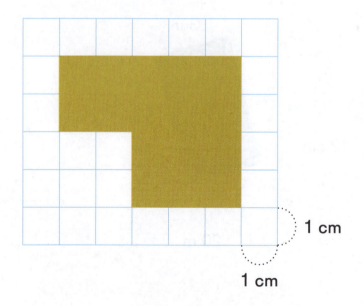

1 cm

1 cm

How can we use rectangles to find the area?

Learn

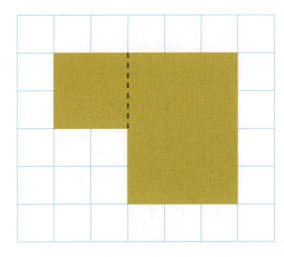

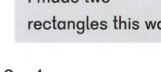
I made two rectangles this way.

$2 \times 2 = 4$

$3 \times 4 = 12$

$4 + 12 =$ | Area = cm²

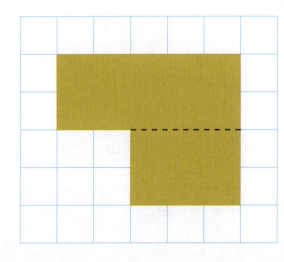

I made two rectangles a different way.

$5 \times 2 = 10$

$3 \times 2 = 6$

$10 + 6 =$ | Area = cm²

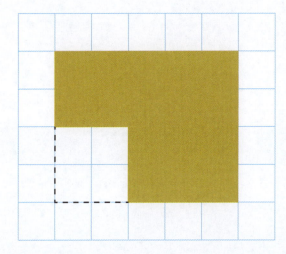

I made a big rectangle first.

$5 \times 4 = 20$

$2 \times 2 = 4$

$20 - 4 =$ | Area = cm²

Do

1 Find the area of each figure.

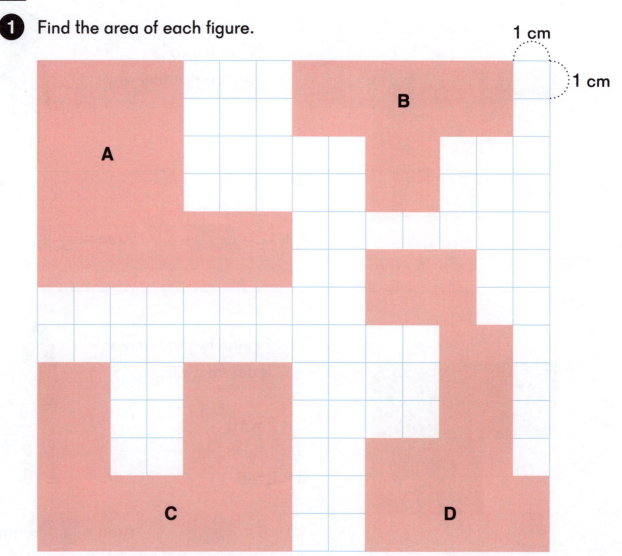

1 cm

1 cm

A

B

C

D

2 Each figure is made up of rectangles.
Find the area of each figure.

(a)

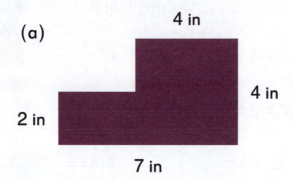

4 in

4 in

2 in

7 in

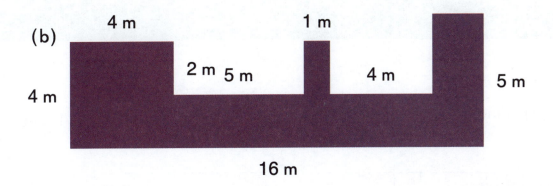

(b)

4 m   1 m

2 m   5 m   4 m   5 m

4 m

16 m

3 Mei has 2 pieces of colored paper.
Each has one side that is 9 in long.
One piece has a side that is 12 in long
and the other has a side that is 5 in long.
She puts them together to make a rectangular poster.
Show two different ways to find the total area.

4 Why do Emma's and Alex's methods for finding the area of this figure work?

Emma's method

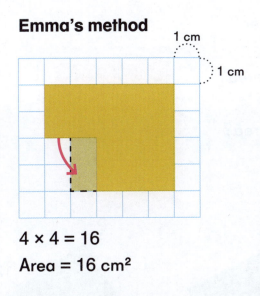

1 cm

1 cm

$4 \times 4 = 16$

Area = 16 cm²

Alex's method

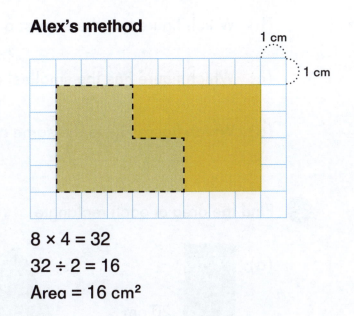

1 cm

1 cm

$8 \times 4 = 32$

$32 \div 2 = 16$

Area = 16 cm²

Exercise 4 • page 153

1 (a) Find the area of each figure.

1 unit
1 unit

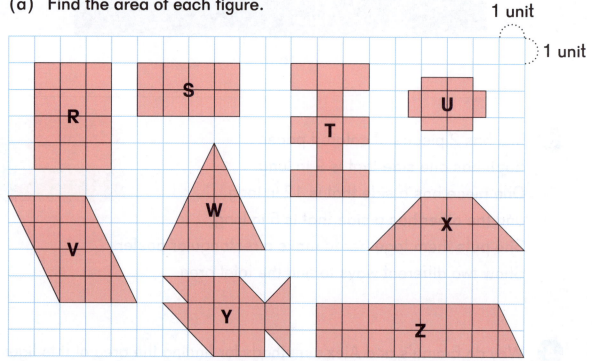

(b) Which figure has the largest area?

(c) Which figure has the smallest area?

(d) Which figures have the same area?

2 Find the area of each rectangle.

(a)

(b)

(c)

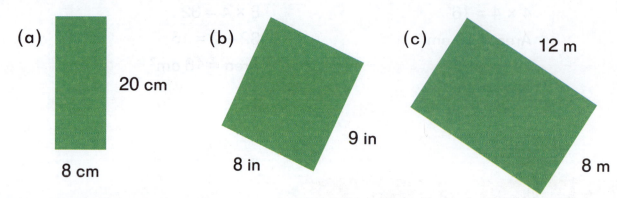

(a) 20 cm, 8 cm

(b) 8 in, 9 in

(c) 12 m, 8 m

3 Find the area of each square.

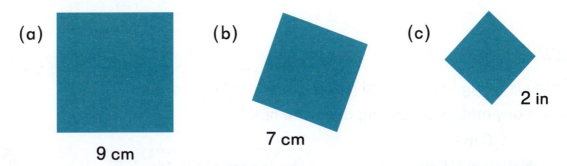

(a)

9 cm

(b)

7 cm

(c)

2 in

4 Find the area of each figure.

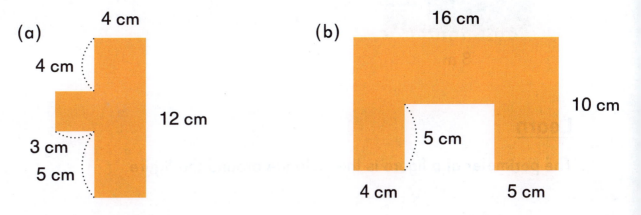

(a)

4 cm

4 cm

12 cm

3 cm

5 cm

(b)

16 cm

10 cm

5 cm

4 cm

5 cm

5 Find the area of this room.

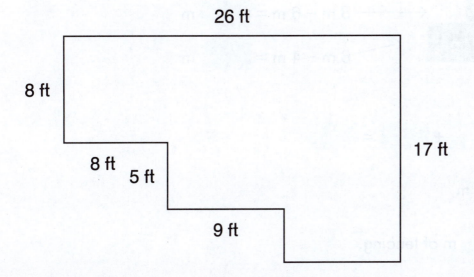

26 ft

8 ft

8 ft

5 ft

9 ft

17 ft

Think

Emma is going to help build a dog run with a fence around it.
How many meters of fencing does she need?

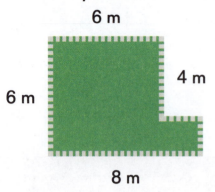

6 m

6 m

4 m

8 m

The length of the fence is the same as the length around the outside of the dog run.

Learn

The **perimeter** of a figure is the distance around the figure.

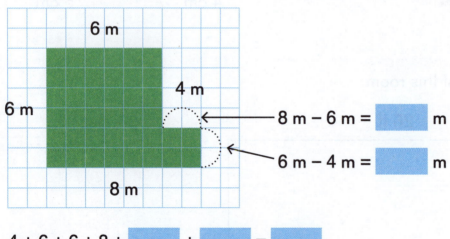

6 m

6 m

4 m

8 m

8 m – 6 m = ⬚ m

6 m – 4 m = ⬚ m

4 + 6 + 6 + 8 + ⬚ + ⬚ = ⬚

Perimeter = ⬚ m

She will need ⬚ m of fencing.

Do

1 Make figures using 12 one-inch square tiles.
What is the perimeter of each figure?

2 Find the perimeter of each figure.

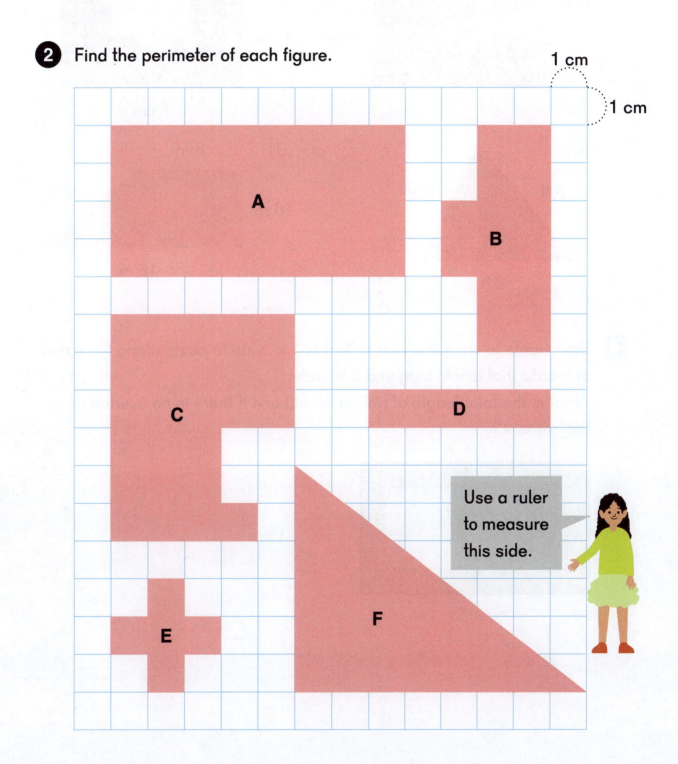

1 cm

1 cm

A

B

C

D

E

F

Use a ruler
to measure
this side.

3 Find the perimeter of each figure.

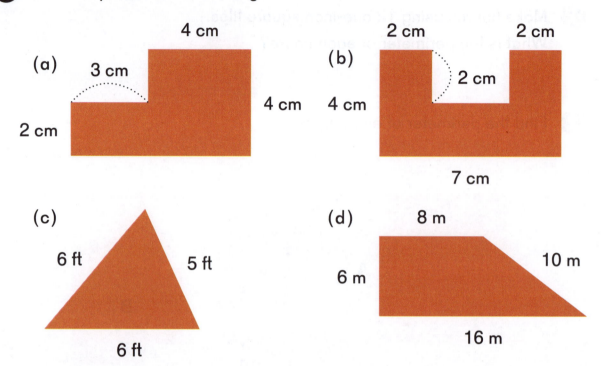

(a)

4 cm
3 cm
4 cm
2 cm

(b)

2 cm   2 cm
2 cm
4 cm   4 cm
7 cm

(c)

6 ft   5 ft
6 ft

(d)

8 m
6 m   10 m
16 m

4 Dion wants to cut some ribbon that is 1 in wide to paste along the sides of a card that is 9 in long and 5 in wide.
What is the total length of ribbon he will use if there is no overlap on the corners?

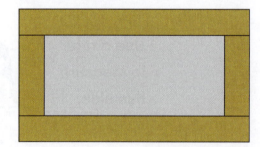

Exercise 6 • page 161

Lesson 7
Perimeter of Rectangles

(7)

Think

Think of ways to calculate the perimeter of the rectangle and square.

5 in

6 in

I wonder if we can use multiplication.

Learn

9 in

5 in  5 in

9 in

$2 \times 9 = 18$ | $2 \times 5 = 10$

$18 + 10 = $

Perimeter = in

---

9 in

5 in  5 in

9 in

$9 + 5 = 14$

$2 \times 14 = $

Perimeter = in

---

6 in

6 in 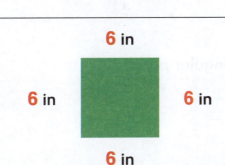 6 in

6 in

$4 \times 6 = $

Perimeter = in

Do

1 Find the perimeter and area of each rectangle.

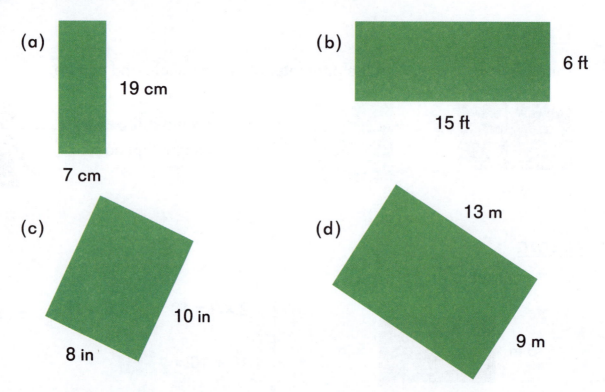

(a) 19 cm, 7 cm

(b) 6 ft, 15 ft

(c) 10 in, 8 in

(d) 13 m, 9 m

2 Find the perimeter and area of each square.

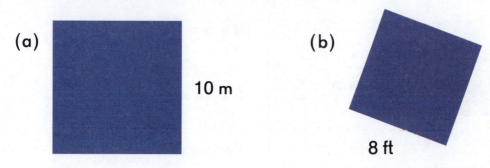

(a) 10 m

(b) 8 ft

3 Sofia walked the distance around a rectangular park.
The park is 56 m long and 29 m wide.
How far did she walk?

Exercise 7 • page 164

Think

Use 12 toothpicks to make rectangles.
What is the area of each rectangle?

Learn

Length = 5 units | Width = 1 unit

Area =   square units

___

Length = 4 units | Width = 2 units

Area =   square units

___

Length = 3 units | Width = 3 units

Area =   square units

Each of these rectangles has a perimeter of 12 units.

Which rectangle has the smallest area?

Which rectangle has the largest area?

Do

1 Use centimeter graph paper to draw all the possible rectangles with side lengths equal to a whole number and an area equal to 12 cm².

(a) What is the longest possible perimeter?

(b) What is the shortest possible perimeter?

2 These figures are made up of one-centimeter squares.

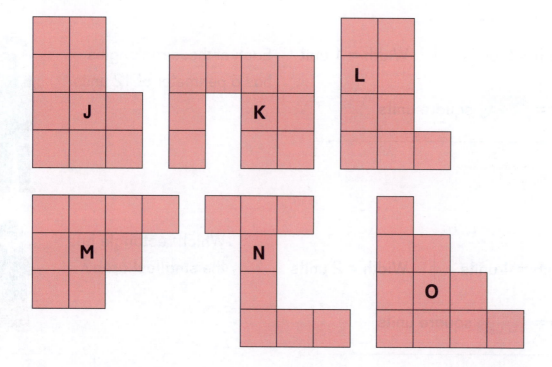

(a) Which figures have the same area but different perimeters?

(b) Which figures have the same perimeters but different areas?

(c) Which figures have the same area and perimeter?

3 Mei and Alex are helping to put a fence around a garden.
They want the garden that has the most space and the least fencing.
Which of the following gardens should they choose and why?

12 m

5 m   **A**

9 m

7 m   **B**

4 Compare the area and perimeter of these figures.
What do you notice?

(a)

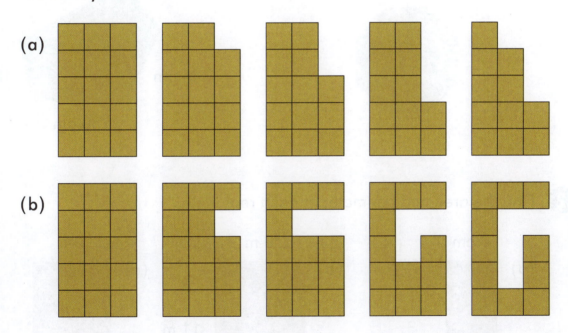

(b)

1 Find the perimeter of each figure.

(a)

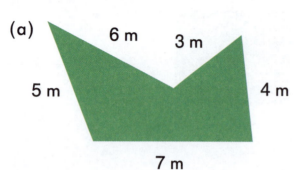

6 m   3 m
5 m   4 m
7 m

(b)

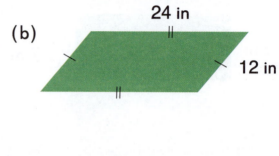

24 in
12 in

(c)

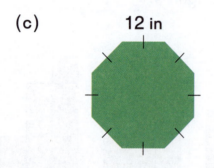

12 in

(d)

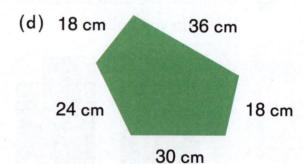

18 cm   36 cm
24 cm   18 cm
30 cm

2 Find the area and perimeter of each rectangle.

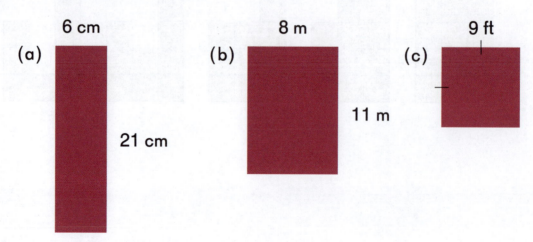

(a) 6 cm, 21 cm

(b) 8 m, 11 m

(c) 9 ft

3 These figures are made from rectangles.
Find the area and perimeter of each figure.

(a)

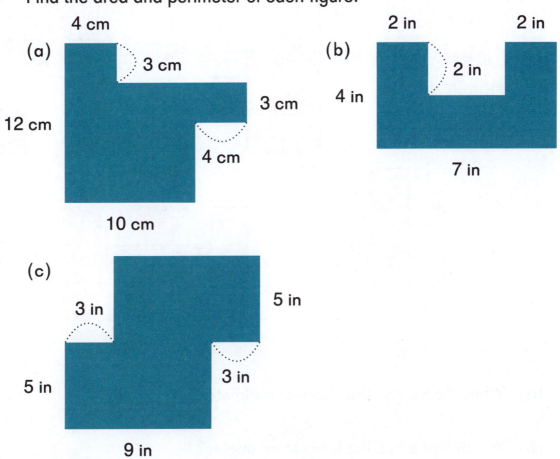

4 cm

3 cm

3 cm

12 cm

4 cm

10 cm

(b)

2 in   2 in

2 in

4 in

7 in

(c)

3 in

5 in

5 in

3 in

3 in

9 in

4 Aki ran around the outside of the football field once.
How far did she run?

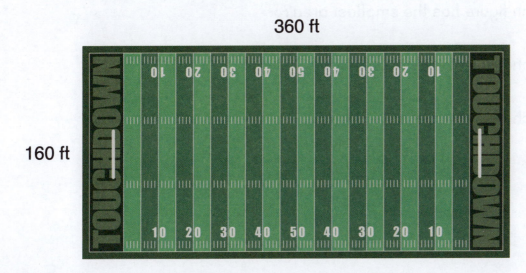

360 ft

160 ft

5 These figures are made up of one-centimeter squares.

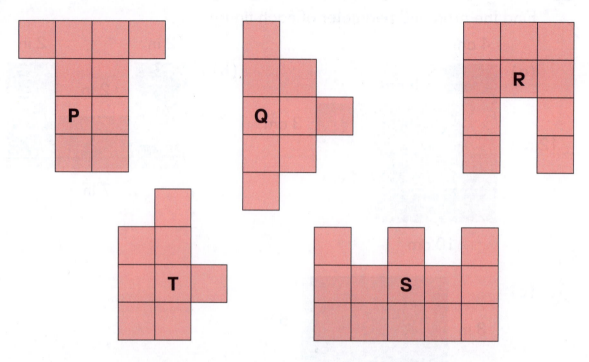

(a) Which figure has the shortest perimeter?

(b) Which figure has the longest perimeter?

(c) Which figure has the largest area?

(d) Which figure has the smallest area?

(e) Which figures have the same perimeter?

(f) Which figures have the same area?

Exercise 9 • page 170

13-9 Practice B

Chapter 14

Time

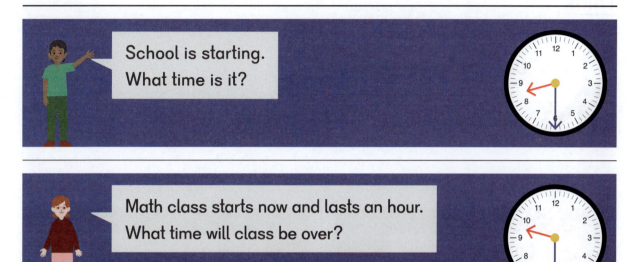

School is starting.
What time is it?

Math class starts now and lasts an hour.
What time will class be over?

Our math game starts now.
How much time do we have to play it?

Lunch is at 12:00.
How much longer until lunch?

Art class ends at 3:00.
How much time do we have
to work on our art project?

Lesson 1
Units of Time

Think

Do something for 1 minute.

I will close my eyes for 1 minute.

I will do jumping jacks for 1 minute.

I will be silent for 1 minute.

What are other ways to measure time?

Learn

Seconds, minutes, hours, and days are units of time.

There are 60 seconds in 1 minute.

There are 60 minutes in 1 hour.

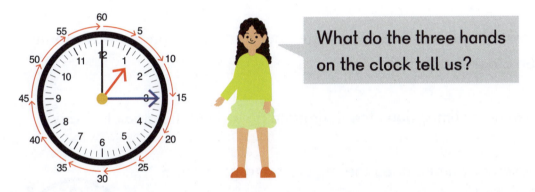

What do the three hands on the clock tell us?

There are 12 hours in the a.m. (morning) and 12 hours in the p.m. (afternoon).

There are [   ] hours in the day.

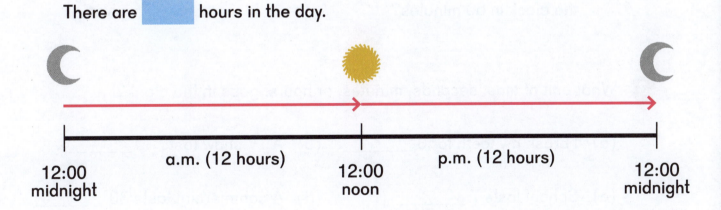

Do

1 Name something that takes about...

(a) 2 s.

(b) 2 min.

We write **s** for seconds,
min for minutes, and **h** for hours.

(c) 2 h.

2 (a) How many minutes pass when the minute hand goes all the way around the clock one time?

(b) How many hours pass when the hour hand goes all the way around the clock one time?

(c) How many times does the hour hand go around the clock in 1 day?

(d) How many times does the minute hand go around the clock in 1 day?

(e) How many times does the second hand go around the clock in 60 minutes?

3 What unit of time, seconds, minutes, or hours, goes in the blank?

(a) I brush my teeth for 3 _____.

(b) A TV show lasts 30 _____.

(c) School lasts 7 _____.

(d) A commercial lasts 30 _____.

4 Dion practiced the violin for 1 h 30 min. Write the time in minutes.

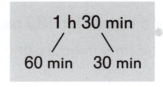

1 h 30 min = [____] min

5 How many minutes are in...

(a) 2 h?

(b) 2 h 5 min?

(c) 3 h?

(d) 3 h 15 min?

6 Sofia took 1 min 50 s to run 400 meters. Write the time in seconds.

1 min 50 s = [____] s

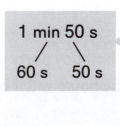

7 How many seconds are in...

(a) 2 min?

(b) 2 min 55 s?

(c) 4 min?

(d) 4 min 32 s?

8 Mei read a book for 130 minutes.
Write the time in hours and minutes.

130 min = [ ] h [ ] min

1 h = 60 min
2 h = 120 min
130 − 120 = ?

9 Write the time in hours and minutes.

(a) 75 min

(b) 100 min

(c) 155 min

(d) 204 min

10 Write the time in minutes and seconds.

(a) 130 s

(b) 110 s

(c) 85 s

(d) 76 s

11 Emma's birthday party will be in 2 days and 4 hours.
How many hours will it be until her birthday party?

2 days 4 hours = [ ] hours

1 day = 24 hours
2 days = ? hours

12 How many hours are in 7 days?

7 days = [ ] hours

13 What goes in the blank, a.m. or p.m.?

(a) The sun set at 7:14 _____.

(b) School starts at 7:40 _____.

(c) The museum opens at 10:00 _____.

(d) Alex went to bed at 8:35 _____.

(e) 1 minute after 12 noon is 12:01 _____.

(f) 1 minute after 12 midnight is 12:01 _____.

We also measure time in weeks, months, and years.

14 Alex's family went on a trip that lasted 3 weeks.
How many days did the trip last?

3 weeks = 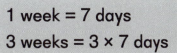 days

1 week = 7 days
3 weeks = 3 × 7 days

Do you know how many days are in each month?

Exercise 1 • page 173

Think

Dion's class left school at 8:50 a.m. for a field trip to the zoo.

(a)   The trip took 50 minutes.
      What time did they arrive?

(b)   The raptor show is at 10:15 a.m.
      How much time passed from when Dion's class arrived at the zoo to the
      time the show started?

(c)   How much time is it from the start of the raptor show to noon?

Learn

(a)

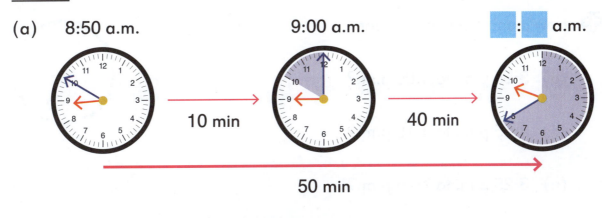

8:50 a.m.   9:00 a.m.   ▢ : ▢ a.m.

10 min   40 min

50 min

They arrived at the zoo at ▢ : ▢ a.m.

(b)

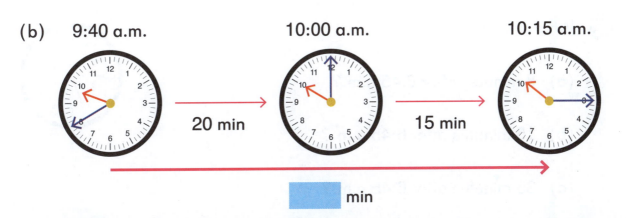

9:40 a.m.   10:00 a.m.   10:15 a.m.

20 min   15 min

▢ min

▢ minutes passed between arriving at the zoo and the raptor show.

(c)

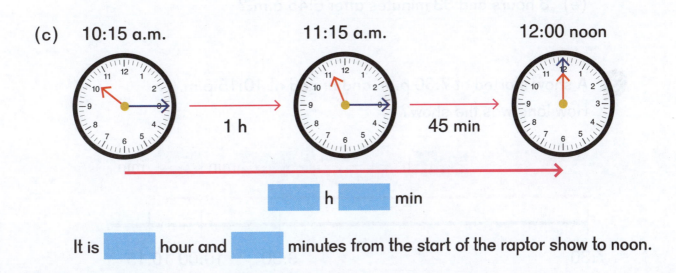

10:15 a.m.   11:15 a.m.   12:00 noon

1 h   45 min

▢ h ▢ min

It is ▢ hour and ▢ minutes from the start of the raptor show to noon.

Do

1 How much time passes from...

(a)   3:25 p.m. to 4:00 p.m.?

(b)   3:25 p.m. to 4:15 p.m.?

(c)   3:25 p.m. to 7:25 p.m.?

(d)   3:25 p.m. to 7:40 p.m.?

2 What time is it...

(a)   8 minutes after 6:45 a.m.?

(b)   15 minutes after 6:45 a.m.?

(c)   35 minutes after 6:45 a.m.?

(d)   3 hours after 6:45 a.m.?

(e)   3 hours and 33 minutes after 6:45 a.m.?

3 A show started at 7:30 p.m. and ended at 10:15 p.m.
How long was the show?

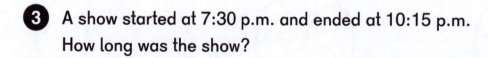

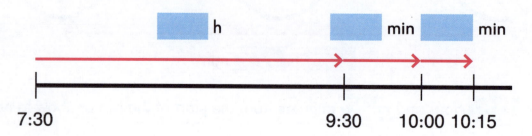

4 Alex went to bed, read for 50 minutes in bed,
then turned off his light and fell asleep 15 minutes later.
How long was it from when he went to bed until he fell asleep?

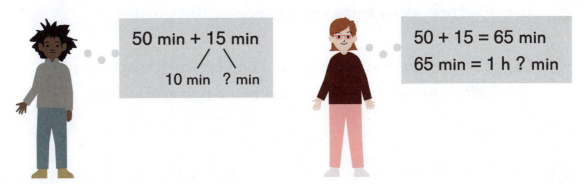

50 min + 15 min
/ \
10 min   ? min

50 + 15 = 65 min
65 min = 1 h ? min

5 Mei went on a hike that took 2 hours and 45 minutes.
The hike started at 8:30 a.m.
When did it end?

8:30 $\xrightarrow{+2\,h}$ 10:30 $\xrightarrow{+45\,min}$ ☐ : ☐ a.m.

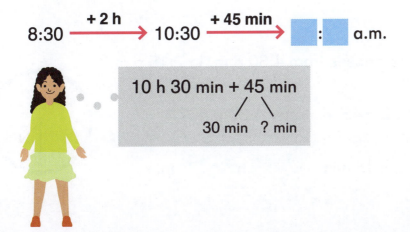

10 h 30 min + 45 min
/ \
30 min   ? min

6 Dion spent 2 hours and 45 minutes reading.
He spent another 1 hour and 20 minutes playing a game.
How much time did he spend on both activities?

2 h 45 min $\xrightarrow{+1\,h}$ ☐ h ☐ min $\xrightarrow{+20\,min}$ ☐ h ☐ min

2 h 45 min + 1 h 20 min = ☐ h ☐ min

Exercise 2 • page 177

1 (a) Find how many times you can hop in 10 seconds.

(b) Using the number of times you can hop in 10 seconds, calculate how many times you could hop in 1 minute.

2 (a) 1 min 45 s = ▢ s   (b) 105 s = ▢ min ▢ s

(c) 2 h 15 min = ▢ min   (d) 125 min = ▢ h ▢ min

(e) 3 days = ▢ h   (f) 7 weeks = ▢ days

3 (a) How much time passes from 4:30 p.m. to 7:15 p.m.?

(b) What time is it 2 h 45 min after 4:30 p.m.?

(c) 4 h 30 min + 2 h 45 min = ▢ h ▢ min

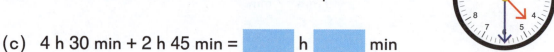

4 (a) 10 s + ▢ s = 1 min   (b) 45 s + ▢ s = 1 min

(c) 35 min + ▢ min = 1 h   (d) 40 min + ▢ min = 1 h

(e) 20 min + ▢ min = 1 h   (f) 6 min + ▢ min = 1 h

5 Add.

(a) 3 h 40 min + 1 h

(b) 3 h 40 min + 20 min

(c) 3 h 40 min + 35 min

(d) 3 h 40 min + 1 h 35 min

(e) 7 h 10 min + 1 h 15 min

(f) 2 h 45 min + 3 h 30 min

(g) 3 h 20 min + 2 h 35 min

(h) 4 h 55 min + 3 h 15 min

6 A baseball game began at 12:25 p.m.
It lasted 2 hours and 28 minutes.
What time did the game end?

7 Mari spent 30 minutes doing math homework, 35 minutes doing reading homework, and 25 minutes doing writing homework.
How much time did she spend doing homework?
Write the answer in hours and minutes.

8 Aaron arrived at the library at 10:25 a.m.
He left the library at 12:00 noon.
How much time did he spend at the library?

9 Natasha left home at 2:20 p.m. and got back at 4:47 p.m.
How long was she gone?

Exercise 3 • page 180

Think

(a) The raptor show ended at 11:20 a.m.
Lunch was 55 minutes later.
What time was lunch?

(b) Dion's class got back to school at 2:25 p.m.
How long were they gone?

Learn

(a)   11:20 a.m.   12:00 noon   ▢ : ▢ p.m.

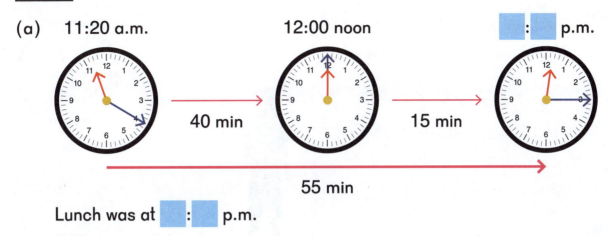

Lunch was at ▢ : ▢ p.m.

(b)   8:50 a.m.

Method 1

Method 2

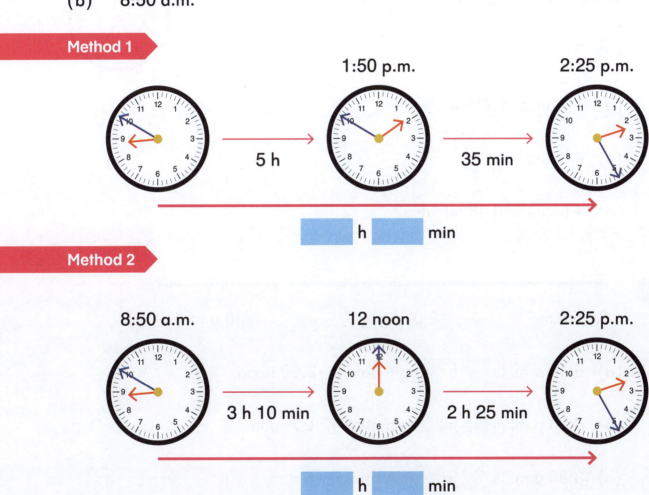

They were gone for ▢ hours and ▢ minutes.

Do

1 The time is 12:45 p.m.
What time will it be in 30 minutes?

> 12:45 → 1:00
> In 15 min, it will be 1:00 p.m.

2 The time is 10:25 p.m.
What time will it be in...

(a) 35 minutes?

(b) 1 hour and 30 minutes?

(c) 4 hours?

(d) 4 hours and 48 minutes?

3

8:45 a.m.   12:00 noon   3:30 p.m.

(a) 8:45 a.m. is ▢ h ▢ min before 12:00 noon.

(b) 3:30 p.m. is ▢ h ▢ min after 12:00 noon.

(c) 3:30 p.m. is ▢ h ▢ min after 8:45 a.m.

4 8 hours and 24 minutes after 8:45 a.m. will be ▢ : ▢ p.m.

5 Sofia was at school from 8:15 a.m. to 3:50 p.m.
How much time did she spend at school?

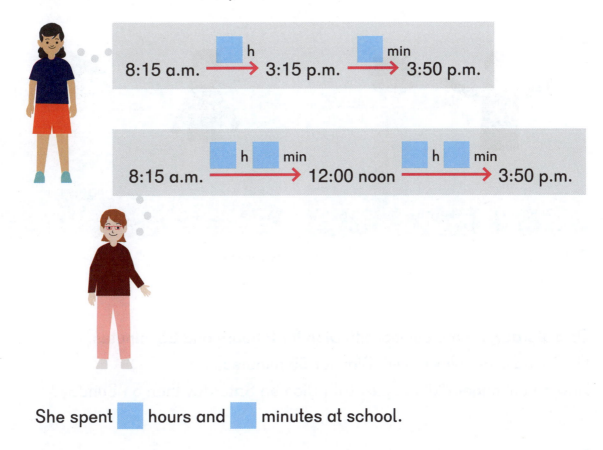

She spent ⬜ hours and ⬜ minutes at school.

6 Alex fell asleep at 9:35 p.m. and slept for 8 hours and 30 minutes.
When did he wake up?

7 Mei's soccer game began at 11:20 a.m.
It ended 2 hours and 27 minutes later.
What time did the game end?

8 A New Year's Eve celebration began at 9:20 p.m.
It ended at 1:05 a.m.
How long was the celebration?

Exercise 4 • page 183

Think

Saturday
2 h 20 min

Sunday
55 min

On Saturday, Emma played with Dion for 2 hours and 20 minutes.

On Sunday, she played with Dion for 55 minutes.

How much longer did she play with Dion on Saturday than on Sunday?

Learn

2 h 20 min − 55 min

Method 1

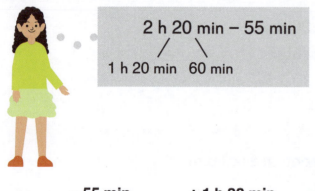

2 h 20 min − 55 min

1 h 20 min   60 min

60 min → **− 55 min** → 5 min → **+ 1 h 20 min** → ▢ h ▢ min

55 min is 5 min less than 1 h.

2 h 20 min $\xrightarrow{-1\,h}$ 1 h 20 min $\xrightarrow{+5\,min}$ [   ] h [   ] min

Method 3

2 h 20 min − 55 min
       / \
  20 min 35 min

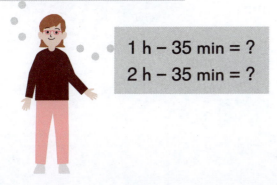

1 h − 35 min = ?
2 h − 35 min = ?

2 h 20 min $\xrightarrow{-20\,min}$ 2 h $\xrightarrow{-35\,min}$ [   ] h [   ] min

Emma played with Dion for [   ] hour and [   ] minutes longer on Saturday than on Sunday.

Do

1 What time is...

(a) 6 minutes before 6:15 p.m.?

(b) 15 minutes before 6:15 p.m.?

(c) 35 minutes before 6:15 p.m.?

(d) 3 hours before 6:15 p.m.?

(e) 3 hours and 30 minutes before 6:15 p.m.?

2 (a) 1 h − 25 min = ▭ min

(b) 3 h − 25 min = ▭ h ▭ min

(c) 3 h 10 min − 25 min = ▭ h ▭ min

(d) 3 h 50 min − 26 min = ▭ h ▭ min

3 Subtract.

(a) 1 h − 20 min | 2 h − 20 min | 2 h 10 min − 20 min

(b) 1 h − 45 min | 6 h − 45 min | 6 h 50 min − 45 min

(c) 1 h − 3 min | 3 h − 3 min | 3 h 50 min − 3 min

4 What time is 3 h 40 min before 7:25 p.m.?

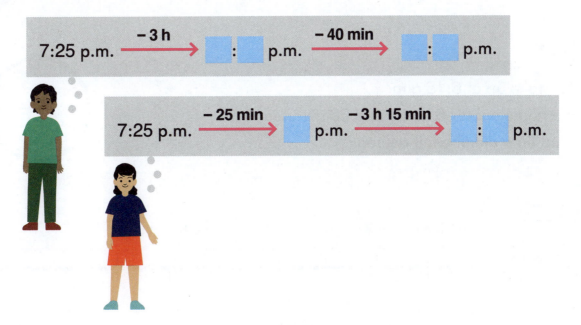

5 (a) 5 h 15 min − 45 min = ⬜ h ⬜ min

(b) 5 h 15 min − 2 h = ⬜ h ⬜ min

(c) 5 h 15 min − 2 h 45 min ⬜ h ⬜ min

6 Subtract.

(a) 1 h 55 min − 1 h 35 min   (b) 3 h 5 min − 1 h 45 min

(c) 6 h 10 min − 1 h 30 min   (d) 5 h 20 min − 1 h 32 min

7 Sofia raked her yard for 1 hour and 50 minutes.
She finished at 4:25 p.m.
What time did she start raking her yard?

Exercise 5 • page 187

Think

Mei woke up at 6:15 a.m.

She had slept for 8 hours and 30 minutes.

What time did she fall asleep?

Learn

Method 1

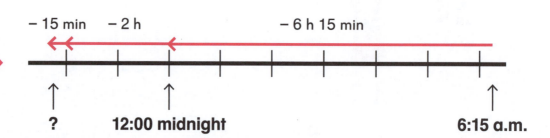

6 h 15 min before 6:15 a.m. is midnight.

2 h before midnight is 10:00 p.m.

15 minutes before 10:00 p.m. is ⬜ : ⬜ p.m.

Method 2

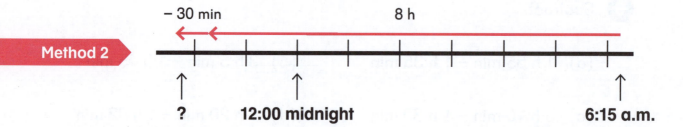

8 h before 6:15 a.m. is 10:15 p.m.

30 min before 10:15 p.m. is ⬜ : ⬜ p.m.

She fell asleep at ⬜ : ⬜ p.m.

Do

1 The time is 1:05 p.m.
What time was it...

(a) 35 minutes earlier?

(b) 1 hour and 5 minutes earlier?

(c) 1 hour and 35 minutes earlier?

(d) 4 hours earlier?

(e) 4 hours and 45 minutes earlier?

2 The time is 5:30 p.m.
What time was it...

(a) 5 hours and 30 minutes earlier?

(b) 6 hours earlier?

(c) 6 hours and 15 minutes earlier?

(d) 12 hours earlier?

(e) 10 hours earlier?

(f) 10 hours and 45 minutes earlier?

3

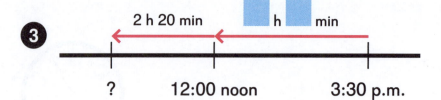

(a) 12 noon is ⬛ hours and ⬛ minutes before 3:30 p.m.

(b) 2 hours and 20 minutes before 12 noon is ⬛ : ⬛ a.m.

(c) What time is 5 hours and 50 minutes before 3:30 p.m.?

4 (a) 20 min before midnight is ⬛ : ⬛ p.m.

(b) 2 h 20 min before midnight is ⬛ : ⬛ p.m.

(c) 2 h 20 min before 1 a.m. is ⬛ : ⬛ p.m.

(d) 4 h before 2:30 p.m. is ⬛ : ⬛ a.m.

(e) 4 h 50 min before 2:30 p.m. is ⬛ : ⬛ a.m.

5 The ferry leaves at 2:30 p.m.
Cars should be in line 1 hour before the departure time.
It takes 2 hours 30 minutes to get from Siti's home to the ferry.
What time should she leave home?

Exercise 6 • page 191

14-6 Calculating Time — Part 4

1 Subtract.

(a) 55 min − 35 min

(b) 12 h − 7 h

(c) 1 h − 35 min

(d) 5 h − 35 min

(e) 5 h − 3 h 35 min

(f) 11 h − 2 h 12 min

2 (a) This clock is 6 minutes fast. What is the correct time?

(b) This clock is 6 minutes slow. What is the correct time?

3 Subtract.

(a) 5 h − 45 min

(b) 5 h 10 min − 45 min

(c) 5 h 10 min − 2 h 45 min

(d) 5 h 55 min − 2 h 45 min

(e) 12 h − 20 min

(f) 12 h 40 min − 20 min

(g) 12 h 40 min − 50 min

(h) 12 h 40 min − 6 h 50 min

(i) 4 h 10 min − 2 h 45 min

(j) 3 h 15 min − 2 h 50 min

4 What time is 3 hours 50 minutes before 1:45 p.m.?

5 Lucia had a 45-minute piano lesson.
The lesson was over at 10:15 a.m.
What time did her lesson start?

6 A baseball game began at 12:25 p.m.
It lasted 2 hours and 28 minutes.
What time did the game end?

7 Andrei arrived at the amusement park at 9:50 a.m.
He left the amusement park at 4:45 p.m.
How long was he there?

8 It took Larry 2 hours and 35 minutes to paint the living room.
It took him another 1 hour and 55 minutes to paint the bedroom.

(a) How long did it take him to paint both rooms?

(b) How much longer did it take him to paint the living room than the bedroom?

9 A school play ended at 1:15 p.m.
The play lasted 2 hours and 30 minutes.
What time did the play start?

Exercise 7 • page 195

14-7 Practice B

Chapter 15

Money

Think

Alex is saving money for a raft.

So far, he has saved these bills and coins.

(a) How much money did he save?
     Write the amount in dollars and cents.

(b) Write the amount in cents.

Learn

(a)

$10 + $5 + $1 = $16

75¢   25¢   12¢

75¢ + 25¢ = $1
$1 + 12¢ = $1.12

$16 + $1.12 = $17.12

$17.12 is read as 17 dollars and 12 cents.
The dot separates dollars from cents.

He saved $⬛⬛⬛.

___

(b) $17.12 = 1,700¢ + 12¢
         = 1,712¢

$1 = 100¢
$10 = 1,000¢
$7 = 700¢

Do

1 Write $2.45 in cents.

$2.45 =  ¢

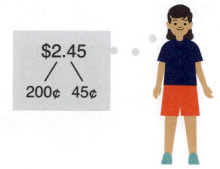

2 Write 460¢ in dollars and cents.

460¢ = $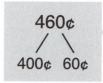

3 Write the amount in dollars and cents and in cents only.

(a)

(b)

4 (a) $23.05 = ☐ ¢

(b) $12.99 = ☐ ¢

(c) $40.90 = ☐ ¢

(d) $0.31 = ☐ ¢

(e) $31 = ☐ ¢

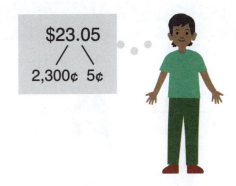

$23.05

2,300¢   5¢

5 (a) 1,543¢ = $ ☐

(b) 820¢ = $ ☐

(c) 2,006¢ = $ ☐

(d) 42¢ = $ ☐

(e) 3¢ = $ ☐

1,543¢

$15   43¢

6 How much more money is needed to make $1?

9 tens and 10 ones make 100.

(a) 85¢   (b) 25¢

(c) 43¢   (d) 71¢

(e) $0.60   (f) $0.35

(g) $0.04   (h) $0.91

Exercise 1 • page 199

Think

Sofia has $7.65.

How much more money does she need to buy a bicycle pump that costs $10.00?

Learn

Add on from $7.65 to make $10.

Method 1

65¢ $\xrightarrow{+\ 35¢}$ $1

$7 $\xrightarrow{+\ \$2}$ $9

65¢ and 35¢ make $1.
$7 and $2 make $9.
$1 and $9 make $10.

Method 2

$7.65 $\xrightarrow{+\ 35¢}$ $8 $\xrightarrow{+\ \$2}$ $10

She needs $ _____ more.

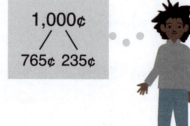

1,000¢
765¢   235¢

Do

1 Match items that together cost $10.

2 How much more money is needed to make $10?
Write the answer in dollars and cents.

(a) 1¢

(b) 85¢

(c) 643¢

(d) $0.60

(e) $9.25

(f) $7.35

(g) $5.71

(h) $6.04

(i) $4.91

Exercise 2 • page 202

Think

$29.40

$9.80

Emma bought a backpack and a thermos.
How much did she spend altogether?

Learn

$29.40 + $9.80

Method 1

$29.40 → **+ $9** → $38.40 → **+ 80¢** → $ ☐

$38.40 + 80¢
         / \
       60¢ 20¢

$29.40 + $9.80
60¢ $9.20

$29.40 —+ 60¢→ $30 —+ $9.20→ $ [　　]

$9.80 is 20¢ less than $10.

$29.40 —+ $10→ $39.40 —− 20¢→ $ [　　]

$29.40 → 2,940¢
$9.80   → + 980¢
            [　　]¢ → $ [　　]

If we line up the dots, we do not have to write the amount as cents only.

Emma spent $ [　　] altogether.

```
  29.40
+  9.80
  [  ].[  ]
```

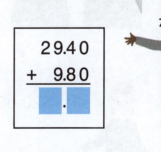

Do

1 Find the value.

(a)   85¢ + 5¢ | 85¢ + 15¢ | 85¢ + 35¢

(b) $6.85 + 15¢ | $6.85 + 35¢ | $46.85 + 35¢

(c) $2.40 + 60¢ | $2.40 + 85¢ | $32.40 + 85¢

2 Find the total cost.

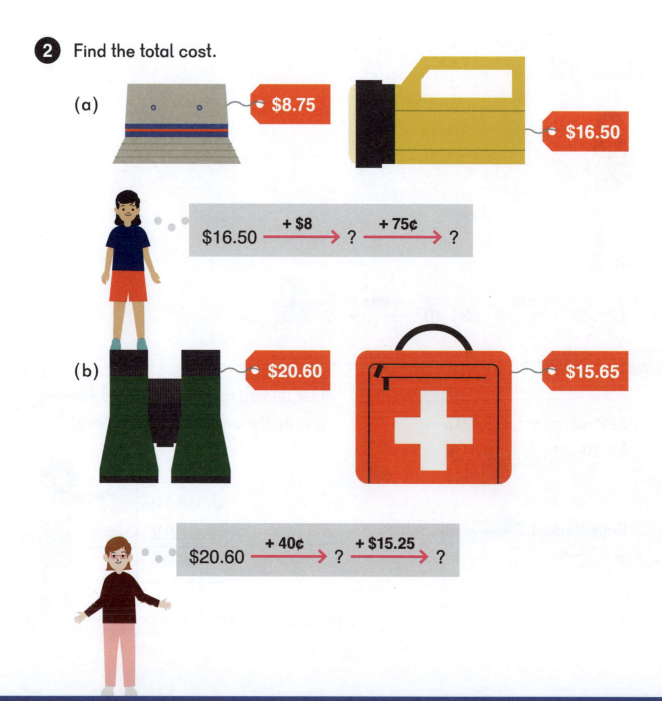

(a) $8.75 $16.50

$16.50 —— + $8 —→ ? —— + 75¢ —→ ?

(b) $20.60 $15.65

$20.60 —— + 40¢ —→ ? —— + $15.25 —→ ?

(c)

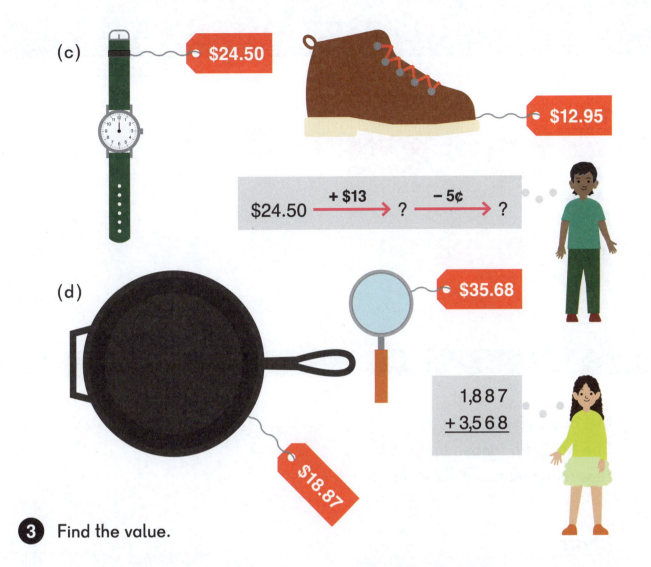

$24.50

$12.95

$24.50 —— **+ \$13** —→ ? —— **− 5¢** —→ ?

$35.68

(d)

$18.87

1,8 8 7
+ 3,5 6 8

---

3 Find the value.

(a) $31.10 + $20

(b) $31.10 + 20¢

(c) $21.75 + $0.25

(d) $21.75 + $0.30

(e) $21.75 + $5

(f) $21.75 + $5.30

(g) $12.70 + $0.60

(h) $15.75 + $2.40

(i) $45.50 + $15.85

(j) $5.09 + $0.92

(k) $13.82 + $7.99

(l) $29.86 + $19.17

Exercise 3 • page 204

Think

$6.85

Mei had $18.50.

She bought a frisbee for $6.85.

How much money does she have left?

Learn

$18.50 − $6.85

Method 1

$18.50 $\xrightarrow{\;-\$6\;}$ $12.50 $\xrightarrow{\;-85¢\;}$ $ [ ]

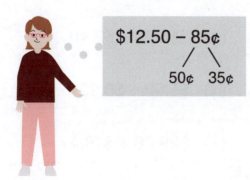

$12.50 − 85¢

50¢   35¢

$18.50 − $6.85
          ╱   ╲
        50¢ $6.35

$18.50 — 50¢ → $18 — $6.35 → $ ▢

Method 3

$6.85 is 15¢ less than $7.

$18.50 — $7 → $11.50 + 15¢ → $ ▢

Method 4

```
  18.50
−  6.85
  ▢.▢
```

```
  1,850 ¢
−   685 ¢
  ▢    ¢
```

She has $ ▢ left.

Do

1 (a) $1 – 40¢ = [ ] ¢ $8 – 40¢ = $[ ]

(b) $1.15 – 40¢ = [ ] ¢ $8.15 – 40¢ = $[ ]

2 Find the value.

(a) $4 – 75¢

(b) $4.50 – 75¢

(c) $13.30 – 65¢

(d) $27.25 – 40¢

(e) $35.05 – 90¢

(f) $50.55 – 60¢

3 Subtract to find the difference in price between the two items.

(a)

$8.75

$16.50

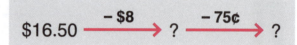

$16.50 —— – $8 —→ ? —— – 75¢ —→ ?

(b)

$20.60

$15.65

$20.60 —— – $15.60 —→ ? —— – 5¢ —→ ?

(c)

$24.50 $\xrightarrow{-\$13}$? $\xrightarrow{+5¢}$?

$12.95

(d)

$$\begin{array}{r} 3{,}568 \\ -\,1{,}887 \\ \hline \end{array}$$

$18.87 $35.68

4 Find the value.

(a) $85.40 – $25

(b) $85.40 – 25¢

(c) $24.15 – $0.85

(d) $75.45 – $16.20

(e) $67.20 – $18.95

(f) $52.01 – $23.75

(g) $42.82 – $6.99

(h) $51.32 – $19.87

Exercise 4 • page 207

Think

Dion bought a bed float for $14.90 and a flamingo float for $18.50.
He paid $40.
How much change did he receive?

$14.90

$18.50

Learn

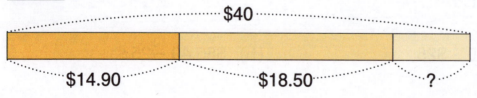

$40

$14.90 $18.50   ?

$14.90 + $18.50 = $ ▮

$40 − $ ▮ = $ ▮

He received $ ▮ change.

I solved it a different way:
$40 − $14.90 = $25.10
$25.10 − $18.50 = ?

Do

1 Emma bought a kite for $15.65.

She has $12.30 left.

How much money did she have at first?

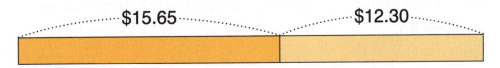

2 A yo-yo costs $24.65.

The jump rope costs $14.85 less than the yo-yo.

How much do the two toys cost altogether?

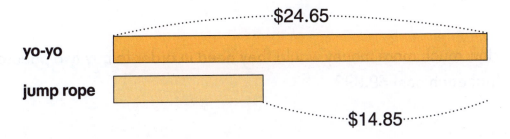

3 Malik saved $8.55 less than Carlos.

Darryl saved $17.75 more than Malik.

Carlos saved $32.40.

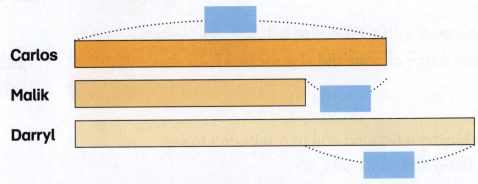

(a) How much money did Darryl save?

(b) How much more money did Darryl save than Carlos?

4

Three people are planning to contribute $32 each to buy a canoe that costs $89.45.

(a) How much money will they have left over?

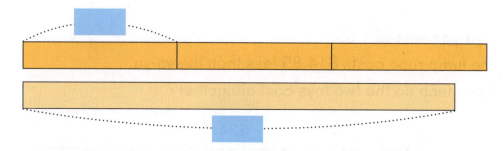

(b) How much more money would they need in order to buy two paddles that each cost $9.99?

5 Mei bought the sunglasses, visor, and ping pong paddle.

(a) How much did she spend altogether?

(b) She paid with a 20-dollar bill. How much change did she receive?

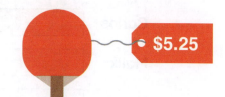

6 Riya is buying a bathing suit and a beach towel.
The bathing suit costs $24.55.
The beach towel costs $12.95 less than the bathing suit.
How much will she pay for the two items?

Exercise 5 • page 210

1 Write the amount in dollars and cents and in cents only.

2 Write in cents.

  (a) $7.05   (b) $18.87   (c) $28.52   (d) $30.20

3 Write in dollars and cents.

  (a) 510¢   (b) 1,204¢   (c) 2,710¢   (d) 3,550¢

4 Find the value.

  (a) $85.80 – $14.10   (b) $13.85 + $36.20

  (c) $28.95 + $17.95   (d) $32.28 + $15.67

  (e) $48.88 – $29.10   (f) $50 – $16.90

  (g) $20.05 – $11.70   (h) $21.92 – $9.58

5. Shanice wants to buy a dress that costs $45.
   She has saved $28.50.
   How much more money does she need to save to buy the dress?

6. Jerome wants to buy a remote control car that usually costs $52.75.
   This week it is on sale for $39.95.
   How much money will he save if he buys it this week?

7. A board game and a hula-hoop together cost $25.60.
   The board game costs $12.95.
   What is the price of the hula-hoop?

8. Dion bought a pail and a shovel.
   He paid with a 10-dollar bill.
   How much change did he receive?

   $5.45

   $2.35

9. After buying a hat for $8.75 and a belt for $7.90,
   Madison had $23.35 left.
   How much money did she have at first?

10. Matt bought 3 t-shirts that each cost the same amount, a pair of pants
    that cost $19.85, and a pair of shoes that cost $24.15.
    He spent $83.
    How much did one shirt cost?

Exercise 6 • page 213

1 (a) ▮ ÷ 9 is 84 with a remainder of 7.

(b) ▮ × 6 = 354

2 What is the greatest even number that can be formed using the digits 7, 8, 4, and 3?

3

(a) What fractions are indicated by P, Q, and R on this number line? Give your answer in simplest form.

(b) Give an equivalent fraction for R with a denominator of 12.

(c) S is how many thirds?

4 (a) Which of the following are greater than $\frac{1}{2}$?

$\frac{5}{6}$ $\frac{3}{5}$ $\frac{3}{10}$ $\frac{8}{9}$ $\frac{3}{8}$ $\frac{9}{8}$

(b) Put the fractions in order from least to greatest.

5 This table and graph show the number of people who visited City Zoo during one week.

Day	Number
Sunday	442
Monday	308
Tuesday	172
Wednesday	489
Thursday	225
Friday	371
Saturday	567

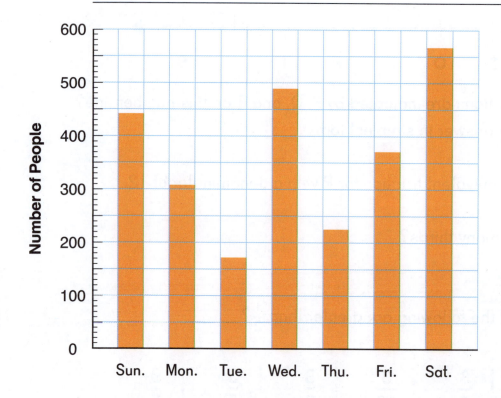

(a) The scale on the left is numbered in increments of ⬜.

(b) Each tick mark shows an increment of ⬜.

(c) Each square on the graph is for an increment of ⬜.

(d) Which day of the week was the most popular for visiting the zoo?

(e) Which day of the week was the least popular for visiting the zoo?

(f) How many more people visited the zoo on Saturday than on Sunday?

(g) Estimate how many people visited the zoo during the weekdays by rounding to the nearest 100.

(h) On the weekend, tickets are $8 each.
How much money did the zoo receive from tickets on Saturday?

(i) On Wednesdays, tickets are $5 each.
How much money did the zoo receive from tickets on Wednesday?

(j) For the rest of the week, tickets are $6 each.
How much less money did the zoo receive from tickets on Tuesday than on Monday?

(k) During the first half-hour on Tuesday, the total ticket sales was $294.
How many people bought tickets during the half-hour?

(l) During the first half-hour on Saturday morning, the total ticket sales was $928 from selling tickets at $8 each.
During the next half-hour, the total ticket sales was $472.
How many more people bought tickets during the first half-hour than the second half-hour?

6   (a) On weekdays, the zoo opens at 10:00 a.m. and closes at 4:30 p.m.
How long is it open each weekday?

     (b) On weekends, the zoo opens 1 h 30 min earlier, and closes at 6:00 p.m.
How long is it open each weekend day?

     (c) At the zoo, Avery bought a toy panda for $12.40 and a t-shirt for $18.95.
She paid with two twenty-dollar bills.
How much change did she receive?

     (d) Avery left the zoo at 3:45 p.m.
She was there for five and a half hours.
When did she arrive at the zoo?

7   One day, a gorilla ate $\frac{5}{8}$ kg of broccoli and $\frac{3}{8}$ kg of kale.

     (a) How many kilograms of both vegetables did it eat?

     (b) How many more kilograms of broccoli did it eat than kale?

8   An elephant at the zoo drank 95 L of water one day.
A giraffe drank 27 L 500 mL that day.
How much more did the elephant drink than the giraffe?

Exercise 7 • page 217

1

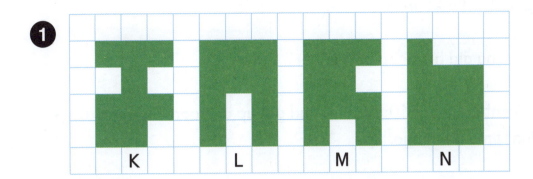

K   L   M   N

(a) Which figure has a different perimeter from the other three figures?

(b) Which figure has a different area from the other three figures?

2 This wall decoration is made from eight triangles.
Each triangle has 3 equal sides.
The perimeter of each triangle is 72 cm.

(a) What is the perimeter of the wall decoration?
Give your answer in meters and centimeters.

(b) Is the wall decoration a rhombus?
Explain why or why not.

(c) What fraction of it is colored green?
Give your answer in simplest form.

(d) Which of the labeled angles are larger than a right angle?

(e) Which of the labeled angles are smaller than a right angle?

3 It took Cooper 15 minutes to fall asleep after he went to bed. He then slept soundly for 4 h 55 min, then was awake for 15 minutes, and then slept again for 3 h 20 min.

(a) How long did he sleep altogether?

(b) He went to bed at 10:30 p.m. and got up as soon as he woke up. What time did he get up?

4 Tina got back from a hike at 2:30 p.m.
She was gone for 4 h 15 min.
What time did she leave?

5 A rectangular field has a length of 120 m and a width of 85 m.
Kawai ran around the field 6 times.
How far did he run?
Give your answer in kilometers and meters.

6 The radius of this circle is 10 cm.
The two shorter sides of the triangle are 12 cm and 16 cm long.
What is the perimeter of this triangle?

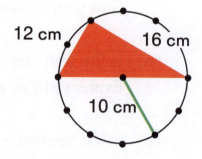

7 Bats cost $24 each and baseballs cost $4 each.
A coach has $200.
After he buys 2 bats, how many baseballs can he buy?

8 A line drawn from which point, B, C, or D, to point A, will form a right angle with this blue line?

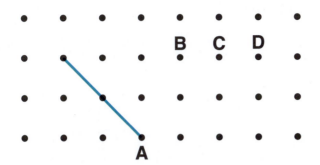

9

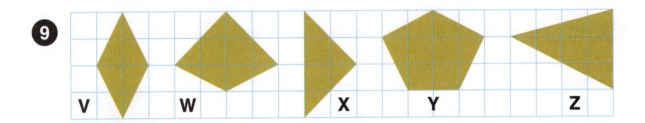

V   W   X   Y   Z

(a) Which figures are quadrilaterals?

(b) Which triangle has two equal angles?

(c) Which figures have a right angle?

(d) Which figure is a rhombus?

(e) Find the area of each figure.

10 The diameter of the largest circle is 112 cm.
What are the radii and diameters of the smaller circles?

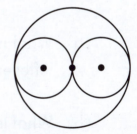

11 When the time is 12:15, do the hands on a clock form a right angle?

12 Carpeting costs $6 per square meter to install.
The carpet itself costs $2 per square meter.
How much will it cost to carpet this room?

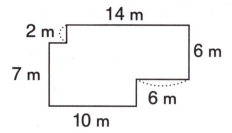

13 This design was made from squares each with sides of 6 cm.

(a) What fraction of the design is blue?
Give your answer in simplest form.

(b) What is the perimeter of the design?

(c) What is the area of the design?

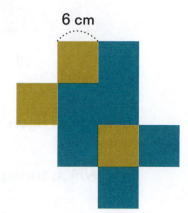

6 cm

14 The length of this glass and wire decoration is 45 centimeters.
The wire used to make the frame around each square piece of
glass was cut from some wire that was 3 meters long.
Each larger square is twice as long as the smaller square.

(a) What is the length of the
leftover piece of wire?

(b) What is the total area of the
glass used in the decoration,
in square centimeters?

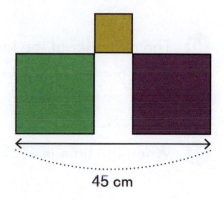

45 cm

Exercise 8 • page 222